OECD Studies on Water

Water and Climate Change Adaptation

POLICIES TO NAVIGATE UNCHARTED WATERS

OECD
BETTER POLICIES FOR BETTER LIVES

This work is published on the responsibility of the Secretary-General of the OECD. The opinions expressed and arguments employed herein do not necessarily reflect the official views of the OECD or of the governments of its member countries or those of the European Union.

This document and any map included herein are without prejudice to the status of or sovereignty over any territory, to the delimitation of international frontiers and boundaries and to the name of any territory, city or area.

Please cite this publication as:
OECD (2013), *Water and Climate Change Adaptation: Policies to Navigate Uncharted Waters*, OECD Studies on Water, OECD Publishing.
http://dx.doi.org/10.1787/9789264200449-en

ISBN 978-92-64-20043-2 (print)
ISBN 978-92-64-20044-9 (PDF)

Series: OECD Studies on Water
ISSN 2224-5073 (print)
ISSN 2224-5081 (online)

European Union
Catalogue number: MN-01-13-045-EN-C (print)
Catalogue number: MN-01-13-045-EN-C (PDF)
ISBN 978-92-79-29937-7 (print)
ISBN 978-92-79-29935-3 (PDF)

IWA Publishing
ISBN: 9781780405742 (Paperback)
ISBN: 9781780405759 (eBook)

The statistical data for Israel are supplied by and under the responsibility of the relevant Israeli authorities. The use of such data by the OECD is without prejudice to the status of the Golan Heights, East Jerusalem and Israeli settlements in the West Bank under the terms of international law.

Photo credits: Cover illustrations: © Taro Yamada/Corbis – © iStockphoto.com/CactuSoup.

Corrigenda to OECD publications may be found on line at: *www.oecd.org/publishing/corrigenda*.

Foreword

Climate change is reshaping the future for freshwater. Water is the predominant means through which the impacts of climate change will be felt. Regardless of future greenhouse gas emissions trends, a certain amount of climate change is already unavoidable. Changes in freshwater systems are already being observed and water risks (shortage, excess, inadequate quality and disruptions to freshwater ecosystems) are expected to increase in a changing climate.

Within this context, the OECD undertook work on water and climate change adaptation to strengthen the evidence base to inform policy responses. It builds on many years of work at OECD on key water issues as well as the economics of climate change adaptation. It also documents trends and draws insight from good practice from the OECD Survey of Policies on Water and Climate Change Adaptation covering all 34 member countries and the European Commission.

This work was overseen by OECD's Working Party on Biodiversity, Water, and Ecosystems (WPBWE) and benefitted from valuable comments and input from delegates. It was authored by Kathleen Dominique with key inputs from Jungah Kim. The risk-based framework was developed by Gérard Bonnis and Kathleen Dominique to inform this work as well as OECD work on water security. The development of the risk-based approach benefitted from a background paper authored by Dustin Garrick and Robert Hope on Economic instruments to manage water security risks.

This work has also benefitted from the comments and guidance of a number of people, including Anthony Cox, Xavier Leflaive, Gérard Bonnis, Michael Mullan, Nicolina Lamhauge, Nick Kingsmill and Lisette Van Marrewijk. The expert review and comments provided by John Matthews of Conservation International and the Alliance for Global Water Adaptation were invaluable. Editorial support from Janine Treves and administrative support from Sama Al Taher Cucci are also gratefully acknowledged.

Table of contents

Country Profiles (**online only**)

Tables

Figures

Preface

Water is the main channel through which the impact of climate change will be felt and the key to developing successful adaptation strategies. We are already observing changes in freshwater systems around the world, and water-related risks – whether there is too little, too much, too unreliable, or of poor quality – are expected to increase. These changes can be either gradual or dramatic, and can jeopardise water security over the long-term making it more costly, as time passes, for governments to adjust to changing circumstances.

We need resolute policy action. The recent *OECD Environmental Outlook* indicates that unless we change our policies and transform our behaviour we are heading towards a world that will be 3.7-5.6 °C warmer than pre-industrial levels by the end of this century. Partly as a result of this by 2050, over 40% of the global population is expected to live in areas characterised as being under "severe water stress". Given these challenges, it is essential that governments adopt a proactive approach to manage the linkages between water and climate change.

Increased water risks and growing uncertainty about future conditions both exacerbate existing water security challenges and complicate any planning, management and investment decision about water. Adapting to new circumstances will require better-informed investment strategies and adaptive water governance that take into account climate variability and minimise potentially costly mismatches between water systems and the future climate.

As awareness of these challenges mounts, governments are stepping up their response. The Netherlands, for instance, has developed "adaptive delta management" to promote flexible strategies for water management and reduce the risk of over- or under-investment in future flood risk management and freshwater supplies. In Australia, water trading allows access to water resources to be reallocated over time in response to changing conditions. In the UK, a "real options" approach has been applied to flood risk management for the Thames Estuary to incorporate the uncertainty of climate change and recognise the value of flexibility into decision making.

This report, *Water and Climate Change Adaptation: Policies to Navigate Uncharted Waters*, provides guidance to policy makers to help improve the prioritisation, efficiency, timeliness and equity of their policy responses. The report sets out a risk-based approach for adapting to climate change that promotes water security. It also documents key trends and highlights best practice from the *OECD Survey of Policies on Water and Climate Change Adaptation*, which covers all 34 OECD countries and the European Commission. This survey is the first of its kind to systematically analyse the challenge posed by climate change for freshwater resources and the emerging policy responses in OECD countries. Finally, the report assesses ways to improve incentives for managing water risks and to increase flexibility in both water policy and financing approaches.

Climate change represents a major challenge for the management of freshwater resources, one that requires a long-term vision and urgent concerted action. I am optimistic that the innovative approach underpinning this report will support governments in their individual and collective efforts to rise to such challenge.

Angel Gurría,

Secretary-General.

Abbreviations

ABI	Association of British Insurers, UK
BMU	Environment Ministry, Germany
CatNat	Insurance Scheme for Natural Catastrophes, France
CCRA	UK Climate Change Risk Assessment
CCRIF	Caribbean Catastrophe Risk Insurance Facility
CLIMATE-ADAPT	European Climate Adaptation Platform
ClimWatAdapt	Climate Adaptation – Modelling Water Scenarios and Sectoral Impacts
CLISP	Climate Change Adaptation by Spatial Planning in the Alpine Space
CONAGUA	National Water Commission, Mexico
CSIRO	Commonwealth Scientific and Industrial Research Organization, Australia
CSOs	Combined Sewer Overflows
DEFRA	Department for Environment, Food and Rural Affairs, UK
DWB	California Emergency Drought Water Bank
EC	European Commission
EPA	Environmental Protection Agency, US
ERDF	European Regional Development Fund
EU	European Union
EU ETS	European Union Emission Allowance Trading System
EUSF	European Union Solidarity Fund
FEMA	Federal Emergency Management Agency, US
GAO	Government Accountability Office, US
GCMs	Global Circulation Models
GEF	Global Environment Facility
IPCC	Intergovernmental Panel on Climate Change
KLIMZUG	Managing Climate Change in Regions for the Future, Germany
LEMA	Law on Water and Aquatic Ecosystems, France
MEDDE	Ministry for Ecology, Sustainable Development and Energy, France
NAS	National Adaptation Strategy
NAP	National Adaptation Plan
NCCARF	National Climate Change Adaptation Research Facility, Australia
NFIP	National Flood Insurance Program, US
NWC	National Water Commission, Australia
NWI	National Water Initiative, Australia
NYC	New York City
OECD	Organisation for Economic Co-operation and Development
ONEMA	National Office of Water and Aquatic Environments, France
PES	Payments for Ecosystem Services

PNACC	National Climate Change Adaptation Plan, Spain
PNUEA	National Plan for Efficient Use of Water, Portugal
RAC	Regional Adaptation Collaborative, Canada
RBMP	River Basin Management Plan
ROD	Record of Decision
SoP	Statement of Principles
TE2100	Thames Estuary 2100
UKCP09	UK Climate Projections 2009
UNFCCC	United Nations Framework Convention on Climate Change
USBOR	US Bureau of Reclamation

Executive summary

Climate change is reshaping the future for freshwater. It aggravates existing strains and complicates future planning, management and investment in water infrastructures. Reducing the adverse consequences and costs of climate change and tapping into any opportunities will require adjusting to new circumstances – that is, adaptation. Adaptation requires flexibility in a domain characterised by long-lived infrastructure with high sunk costs. It requires foresight where there is low confidence in climate projections for key water parameters at local scale. It also calls for adaptive water governance where inertia and poor water governance are more often the norm than the exception. In response to this challenge, OECD countries are making progress tackling this issue and a number of lessons learned can be drawn from experience to date.

Climate change is, to a large extent, water change. Water is the predominant means through which the impacts of climate change will be felt. More torrential rains, floods and droughts can be expected in many areas. Changing precipitation patterns are shifting rainy seasons and affecting the timing and quantity of melt water from snow pack and glaciers. In many cases, these impacts are making flood protection, water storage, urban drainage, water supply and treatment more costly. Shifts in extremes are likely to create a bigger challenge for adaptation than shifts in averages. They are also likely to be the key cost drivers for adapting water infrastructures. Freshwater ecosystems and the services they provide are especially vulnerable.

Despite abundant evidence of climate change impacts on freshwater, there are **significant gaps in the existing evidence base** that pose challenges for informing practical, site-specific adaptation. The level of confidence in climate change projections for key water parameters decreases as their potential utility for adaptation decision-making increases. However, one trend appears predictable: **the future for freshwater will not look like the past.**

A risk-based approach to adaptation

A **risk-based approach** can provide a flexible, dynamic, and future-oriented approach to adaptation in the absence of reliable climate predictions. A risk-based approach encourages policy makers to consider a whole range of possible future conditions, from the commonplace to the extremely unlikely and weigh the alternatives. Adaptation should not be undertaken in a way that focuses only on climate as a risk driver to the exclusion of other, often more dominant, drivers of water risks, such as social, economic and political systems. At the same time, adaptation should be seen as a prerequisite to improving water security over the long-term.

There is a need to "know", "target" and "manage" water risks. "Knowing" the risk requires the incorporation of both scientific and technical inputs into risk assessments as well as risk perceptions. Even under pervasive uncertainty, a range of decision-making approaches (e.g. sensitivity analysis, scenario-based approaches) can be used to help "know" the risk. "Targeting" the risk requires determining the *acceptability* of water risks, in light of potential consequences and costs of amelioration. This also requires weighing "risk-risk trade-offs" that can arise when efforts to reduce a given risk such as shortage may increase other risks, such as disrupting the resilience of freshwater systems. "Managing" the risk requires clarity in terms of risk sharing arrangements between public and private actors. It also means considering all risk management strategies (avoid, reduce, bear or transfer) and applying the full range of policy instruments available. **Policy guidelines** can help to prioritise action and improve the timeliness, efficiency and equity of responses.

Progress on adaptation for water in OECD countries

In order to gauge progress and gain insights from practical adaptation efforts for water systems, the OECD Secretariat undertook a **survey of policies across all 34 member countries and the European Commission**. The results reveal general trends and lessons learned. The survey documents that water is not just an important "sector" for adaptation, but that it is also **an essential resource, as well as a potential threat**, affecting a number of other policy domains – energy, agriculture, infrastructure, biodiversity, and health. Just as energy is a key for mitigation, water is the key for adaptation.

The survey reveals that **all** 34 OECD **countries have already observed changes in freshwater systems**. Nearly all countries **project increasing water risks** due to climate change. **Extreme events** (e.g. floods and/or droughts) are cited as a primary concern by 32 countries, along with the European Commission (EC). **Water shortage** is a key issue for 23 countries, as well as the EC. **Water quality** is a key concern for 15 countries, while impacts on **water supply and sanitation** were flagged by 16 countries. For 13 countries, **freshwater ecosystems** were among their primary concerns.

The majority of efforts to date in OECD countries have focussed on "**knowing the risk**" by building the scientific evidence base and disseminating information. Governments should ensure that this evidence is used to best effect and meets the needs of users in making practical, on-site adaptation decisions. In terms of **policy instruments**, information-based instruments (e.g. flood risk maps, decisions support tools for risk management, adaptation guidance for local governments) are by far the most widely used.

To "target" and "manage" the risk, some countries are revising **laws and regulations**, such as sustainable water abstraction limits, building codes and land-use planning. Other measures include adjusting **economic instruments** (e.g. water tariffs, water-related environmental taxes, flood insurance schemes) to reduce baseline stress on water systems, raise financing and address flood risks. Only a handful of countries have begun to explicitly address the issue of **financing**.

Laying the groundwork today to prepare for the future

Adaptive water governance and sound water policy will go a long way to enhancing resilience to climate change. However, most water policy instruments have not been specifically designed with climate change adaptation in mind and may need to be adjusted in light of new evidence. At the same time, some existing policy settings may undermine effective and efficient adaptation by distorting market signals. Climate change strengthens the case for addressing these inefficiencies.

Scaling up the use of **economic instruments** and/or reforming perverse incentives (such as subsidising water supply for certain users, encouraging development in areas at high risk of flood) can provide flexibility and minimise the costs of adjusting to changing conditions. **Flood insurance** can provide incentives to reduce exposure and vulnerability to floods, efficiently spread residual risk and offset the economic impact of water-related disasters. However, greater uncertainty about future climatic conditions makes appropriately pricing flood insurance increasingly difficult. **Water trading** can improve the efficiency of water allocation in response to changing conditions. Temporary transfers can be effective for managing short-term supply variability, but, on their own, they are not sufficient to adjust to long-term declines in water availability. **Efficient water pricing** can boost adaptation by reducing inefficient water use, encouraging the diversification of sources of supply and raising financing for potentially higher investment needs. Incentives for **ecosystem-based adaptation** and **green infrastructure** can provide a cost-effective means to address uncertainty by avoiding or delaying lock-in to capital-intensive infrastructure and provide an additional option value, as compared to alternatives.

Climate change raises new issues for **financing** water and will likely add to the existing funding shortfall. In general, financing adaptation should build on sound approaches to financing water systems generally and avoid skewing financing to "speciality" projects that might be easily labelled as adaptation, but do not necessarily maximise net benefits. Countries should also focus on spending available financing wisely by using flexible investment approaches, such as **real options**, to deal with uncertainty.

Long-term climate change is a novel challenge that will test conventional approaches. The case for action is well-documented and compelling. Not all water risks can be avoided, but well-prepared, **resilient water systems will be better able to adjust to new conditions, at lower cost, and bounce back from disasters more quickly**. Yet, experience shows that mounting scientific evidence and a robust economic case do not necessarily spur action. History is full of examples from water crises and lessons for how they could have been avoided or better managed. Indeed, hindsight can provide valuable guidance for climate change adaptation and help to prepare for the future.

Reforming water systems takes time, stakeholder engagement, and political will, so action needs to start now. Countries that shore up their water governance today and put in place the policies needed to prepare for the future can avoid managing water crises in a reactive and more costly way.

Chapter 1

A changing and uncertain future for freshwater

This chapter provides an overview of the range of complex impacts on freshwater that can be expected in a changing climate. It also highlights the main sources of uncertainty and information gaps associated with climate change impact assessments on water systems that pose challenges for informing practical, on-site adaptation decisions. Finally, it examines the policy implications of the proposition that "stationarity is dead", or in other words, that the future for freshwater will not look like the past.

Key messages

- **Climate change is, to a large extent, water change.** Climate change affects all aspects of the water cycle and water is the main way through which the impacts of climate change will be felt. The consequence of these impacts will depend on their nature, where and when they occur, and the exposure and vulnerability of the populations, ecosystems, and physical assets they affect.
- Typical assessments of the impacts of climate change on freshwater are of **limited use when it comes to making practical, on-site decisions** about adaptation. In general, the level of confidence in climate change projections decreases as their potential utility for making decisions on how to adapt increases. Adaptation decisions need to accommodate considerable uncertainty.
- One trend appears predictable: **the future for freshwater will not look like the past.** This shift calls for a flexible, dynamic, future-oriented approach that takes into account climate variability on all timescales.

Climate change impacts on freshwater resources are already evident and are projected to become more significant and to accelerate over time (Bates et al., 2008). There is also a growing recognition that climate change presents a singular challenge for water systems[1] by rendering the historical assumption of stationarity[2] increasingly irrelevant (Milly et al., 2008). This means that a fundamental assumption upon which water management, infrastructure design and planning, and ultimately many economic and resource management decisions are founded will no longer be a reliable basis for future planning and management. Decisions made today may lock us into management strategies and infrastructure for many decades that will not match future climatic conditions. The unprecedented rate of change and potential novel changes outside of historical experience introduce a greater degree of uncertainty beyond what water managers have traditionally had to cope with.

Uncertainty and knowledge gaps

Despite an ever-expanding scientific basis, reliable information about the nature, magnitude and timing of hydrological impacts at the scale needed for water resources planning and management is generally lacking. Improving this information base to attain the required level of detail and confidence needed to inform practical, on-site adaptation decisions will take time. This is due to limitations in data, modelling capacity, and computational requirements. Moreover, by their very nature, climate and hydrological systems are hard to predict. Given the current state of knowledge and limits to predictability of climate change impacts on water, effective decisions to adapt to climate change will need to be made in the absence of accurate and precise climate predictions (Dessai et al., 2009).

Most climate change impact studies rely on projections from **global circulation models** (GCMs). These models were originally designed to assess the global impact of various emissions pathways in order to make the case for mitigation efforts. The extension of their use to adaptation decision-making is a relatively recent development. The current suite of climate models were not developed to provide the level of accuracy required to inform adaptation decisions for water resources management (Kundzewicz and Stakhiv, 2010). The utility of GCMs for adaptation decisions for water resources and the most promising approaches for addressing their shortcomings or developing alternatives are the subject of widespread debate (Kundzewicz and Stakhiv, 2010). Even if some of the limitations of current approaches can be addressed, adaptation decisions will still need to accommodate considerable uncertainty (Anagnostopoulos et al., 2010; Wilby, 2010; Bates et al., 2008).

Although is it becoming "standard" practice for climate change impact assessments for water to link the results of a climate change model for temperature and precipitation with a hydrological model for runoff, these assessments have several limitations (Rodríguez-Iturbe and Valdés, 2011). However, the emergence of a "standard" approach does not imply that there is consensus about the utility or effectiveness of this approach. Novel approaches, such as "decision scaling" (Brown and Wilby, 2012), are emerging as a promising way forward (see Chapter 2 for further discussion). The main sources of uncertainty and knowledge gaps associated with typical climate change impact assessments for water resources are highlighted below.

Uncertainty related to scenarios, emissions trends, and models

There is significant uncertainty associated with climate models' reproduction of the current climate and simulation of the future climate. These models simulate some climatic processes in only a rudimentary fashion (Bates et al., 2008). Other sources of uncertainty are the scenarios used to estimate emission trends and the way that climate models simulate the impact of those trends on the climate. Depending on the climate model, the same emission trend can produce a wide variation of climate change projections.[3] This is particularly acute in the case of precipitation and evapotranspiration. Hydrological models also add substantial uncertainty due to regional differences and limitations in the coverage of monitoring networks (Huntington, 2006).

Modelling uncertainty may be reduced to some extent by running an ensemble (several slightly different models of the climate system) or thousands of runs from a single model. Yet, in practice, this is a complex and resource-intensive task. Also, the results from an ensemble run may diverge significantly. Even multiple runs of a single model can show significant variation between projections at coarse spatial and temporal scales. Lack of agreement between climate models does not mean that there will be no impact, or that any given impact is unlikely. Instead, it may mean that there is a large range of possible futures, including significant potential increases or decreases in a given climate parameter.

Coarse resolution/scale mismatch

In general, the level of confidence in climate change projections decreases as their potential utility for adaptation decision-making increases. As the spatial scale decreases, projections become less consistent between models. Thus, it is widely recognised that findings from global assessments of climate and hydrological change are not directly usable by decision makers at regional, national and subnational levels for adaptation (UNFCCC, 2011).

The coarse "resolution" of global climate models means that outputs are insufficiently detailed for climate impact studies at finer geographic scales. While climate models have been able to reproduce broad features of the past climate at large geographic scales (continental and above), in general, they still cannot reconstruct the important details of the climate at finer scales (Kundzewicz and Stakhiv, 2010). Because of these limitations, outputs from GCMs are typically "downscaled". These techniques require significant information on the ground for calibration (Rodríguez-Iturbe and Valdés, 2011). However, efforts to address the scale mismatch of global models to provide more site specific information through downscaling also have serious practical limitations (Wilby and Dessai, 2010).

Low confidence for key climate parameters

There is a higher degree of confidence in estimates of shifts in temperature, than for changes in precipitation and evapotranspiration. Unfortunately, precipitation, the principal input to freshwater systems, is not adequately simulated in present climate models (Kundzewicz et al., 2008). With some exceptions, models generally disagree about the magnitude of precipitation changes and sometimes also the direction of the change. Low confidence in precipitation projections also precludes the reliable estimate of changes in flood frequency and magnitude.

A focus on shifts in the mean

Changes in averages are easier to project than changes in extremes. However, projections of increasing average temperatures and changes in average annual rainfall are of limited use for adaptation decision-making. Reliable estimates of changes in the forms of precipitation (rain or snow), seasonal timing of precipitation, inter annual variability, shifts in runoff and river discharge are generally lacking. Moreover, most models do not even try to simulate shifts in extremes, often assuming that current levels of variability will continue relative to a shifting mean. This assumption may overlook some of the most severe, sudden, and costly impacts of climate change on water. Figure 1.1 illustrates several modes of climate change. Climate models tend to focus on changes in the "mean", which is not likely to be either the most likely mode of change (Le Quesne et al., 2010), nor the most relevant for adaptation decisions.

Figure 1.1. **Modes of climate change**

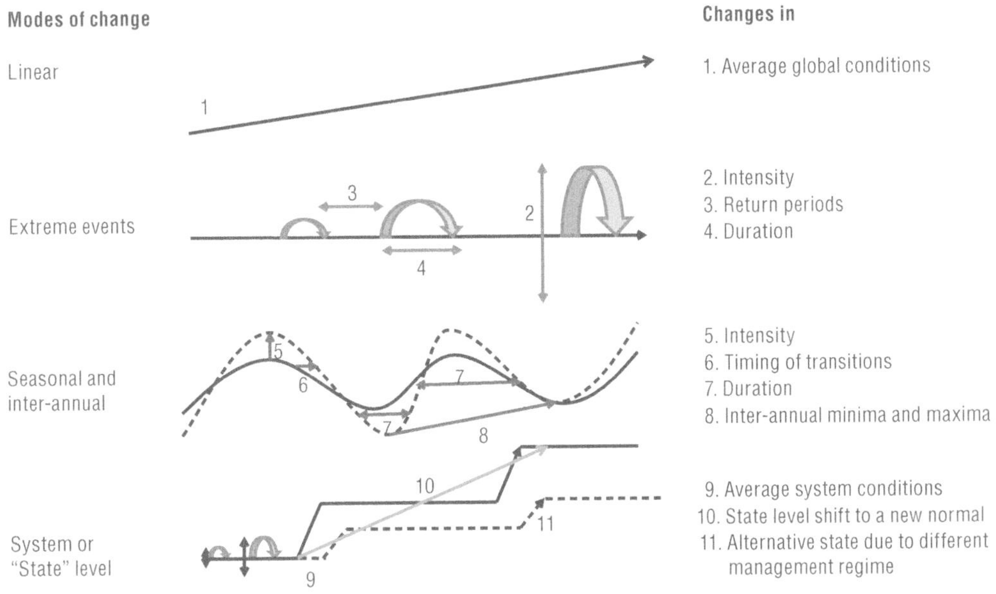

Source: A.J. and J.H. Matthews (2012), Adapted from "Vulnerability to What Change?", presented at the UNFCCC Technical workshop on water, climate change impacts and adaptation strategies, Mexico City, Mexico 18-20 July, World Wildlife Fund and Conservation International, *http://unfccc.int/adaptation/workshops_meetings/nairobi_work_programme/items/6955.php* (accessed 12 December 2012).

A widening range of possible futures

Uncertainties in climate change impact assessments for water resources arise in all stages of the impact assessment process. Figure 1.2 illustrates how the sources of uncertainties compound to multiply the range of possible futures.

Scientific advances can either reduce or expand the range of uncertainty. For example, as previously unknown processes are identified and described uncertainties about those processes may expand the existing range of uncertainty. Even if perfect climate models could be built, uncertainty about all the other non-climatic pressures means that regional hydrological projections would still be highly uncertain (Wilby, 2010).

Figure 1.2. **Cascade of uncertainty**

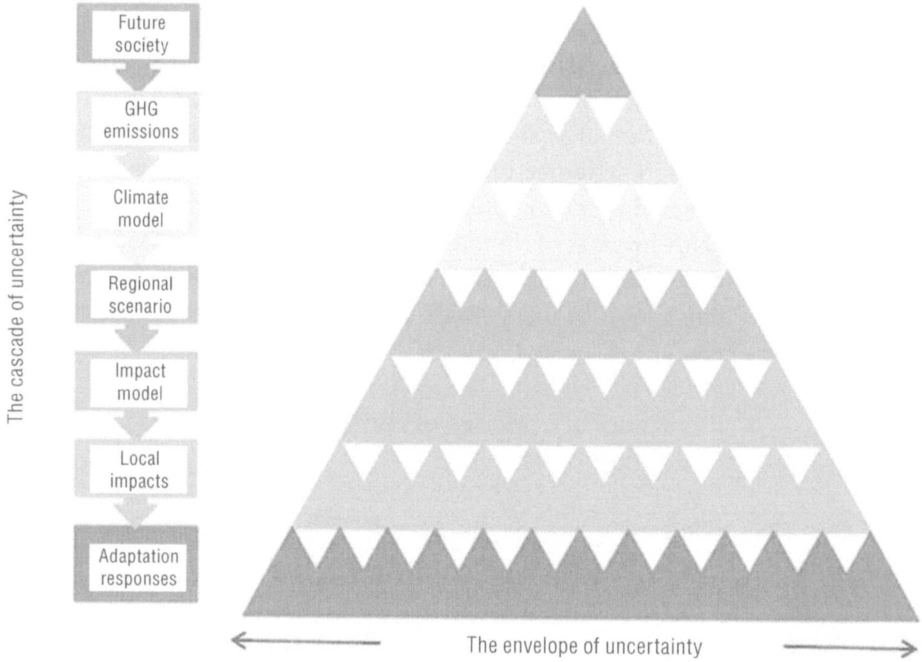

A cascade of uncertainty proceeds from different socio-economic and demographic pathways, their translation into concentrations of atmospheric greenhouse gas (GHG) concentrations, expressed climate outcomes in global and regional models, translation into local impacts on human and natural systems and adaptation responses.
Source: Wilby, R.L. and S. Dessai (2010), "Robust Adaptation to Climate Change", *Weather*, Vol. 65/7, Royal Meteorological Society, Reading, pp. 180-185, *http://dx.doi.org/10.1002/wea.543*.

A future for freshwater unlike the past

Despite all of the uncertainty, one trend appears predictable: the future will not look like the past. The notion that "stationarity is dead" (Milly et al., 2008) and is no longer an adequate guide for future water resources risk assessment and planning has gained widespread acceptance (Wilby, 2010). This marks a significant departure from the past, as much of the experience to date in managing water resources and infrastructure is based on the historical record of climate variability during a period of relatively stable climate.

Stationarity has served as a central, default assumption for water resources management. Probability-based design using historical climate data informs the construction and operation of levees, dams, spillways, and water supply and sewage treatment systems. Flood frequency characteristics, although meaningful under the assumption of stationarity, are questionable in the nonstationary environment (Kundzewicz and Somlyody, 1997). Water allocation regimes are oftentimes based on historical climate data, as are the basic tools that the insurance industry uses to communicate about water risks. Climate change alters the assumptions, data and modelling techniques required to develop information about water risks, such as flood zone maps and, ultimately, their utility as the basis for insurance (Ludwig and Monech, 2009).

The declining relevance of historical climate information to inform current and future planning presents a major challenge for water managers and policy makers. This shift signals the end of the static design paradigm for water resources systems, in favour of a dynamic response to changing conditions at all timescales (Brown, 2010). Such a dynamic approach would consist of periodically adjusting forecasts, building a flexible system focussed on robustness rather than optimisation, and taking a multidisciplinary approach to managing risk. A flexible, dynamic approach is required to minimise potential mismatches between water infrastructures and future climate (Matthews et al., 2011).

Climate change as water change

Despite the uncertainty and knowledge gaps, there is a significant and growing body of scientific evidence documenting the range of complex changes in the water cycle that can be expected in a changing climate. This evidence is useful to provide an overview of the broad range of changes that will present challenges for managing water systems in the future.

Climate change affects all aspects of the water cycle and water is the predominant means through which the impacts of climate change will be felt. Climate change is driving an on-going intensification (increases in evapotranspiration and precipitation) of the water cycle[4] (Huntington, 2006). In other words, in a warmer atmosphere, the water cycle is speeding up.

A certain amount of climate change is already unavoidable, regardless of future greenhouse gas emissions. Significant impacts have already occurred. The OECD *Environmental Outlook to 2050* projects that without more ambitious mitigation policies, the mean global temperature could increase by 3 °C to 6 °C above pre-industrial levels by the end of the century (OECD, 2012). Significant action is required to limit mean increases to 2 °C. Even if this target is achieved, it will only serve to slow the rate of climate change and there will still be considerable impacts.

Climate change impacts on freshwater include shifts in precipitation patterns, rising water temperature, deteriorating water quality, increases in evapotranspiration, and increases in the frequency and intensity of extreme events (Bates et al., 2008; IPCC, 2007). Impacts are expected to become more pronounced over time and the rate of change is expected to accelerate, with more severe impacts anticipated in the second half of the century. The range of impacts on water are summarised below.

Changing precipitation patterns and increased variability

Climate change is projected to shift the spatial and temporal distribution of precipitation, with some regions becoming wetter, others drier. In general, regions with high rainfall are projected to receive more precipitation, while arid and semi-arid regions are projected to become drier. Shifting precipitation patterns will affect runoff (see below), the rate of surface and groundwater recharge, and displace rainy seasons. More frequent and intense precipitation increases erosion and sediment loads in rivers, lakes and coastal zones with a negative effect on water quality. In arid and semi-arid regions, any reduction in rainfall has serious implications for rivers and lakes, even causing them to dry up, as seen in the case of Lake Chad (Ludwig and Moench, 2010).

Over the past several decades, increases in precipitation have been observed over land in high northern latitudes, while decreases have dominated in areas situated between 10 °S to 30 °N. Over the same period, land classified as very dry has more than doubled globally (Bates et al., 2008). Overall, the attribution of observed changes in global precipitation to climate change remains uncertain because precipitation is strongly influenced by large-scale patterns of natural variability.

Projections indicate increases in precipitation, average river runoff and water availability in high latitudes and in some parts of the tropics. Some dry regions at mid-latitudes and in the dry tropics are projected to become drier (Bates et al., 2008). Many semi-arid and arid areas (e.g. Mediterranean basin, western USA, southern Africa, and north-eastern Brazil) are particularly vulnerable to the impacts of climate change and are projected to suffer a decrease in water resources (IPCC, 2007). Higher temperatures will also alter the proportion of precipitation falling as rain and snow. The form of precipitation is extremely important for snowpack-dominated regions (e.g. the Sierra Nevadas, the Andes and the Himalayas). These areas depend on seasonal snowpack to meet water demand in dry seasons. The destruction of natural storage in the form of snowpack means that alternative storage will be required to ensure that winter precipitation continues to be an economically valuable resource during the summer when demand is high.

Changes in precipitation manifest at regional or local scales and are generally poorly described by climate models. These changes are among the least well-understood and least predictable aspects of climate change while at the same time, they are among the most important for making adaptation decisions for water resources management.

Shifts in runoff and river discharge

One of the most significant climate change impacts for water resources is the change in river discharge (Ludwig and Moench, 2010). Low river flows affect the functioning of freshwater ecosystems, navigation, hydropower generation, the availability and quality of water supply.

Observed changes in runoff and river discharge trends are not always consistent with changes in precipitation, although this may be due to data limitations. At the global scale, several studies suggest changes in annual runoff, with some regions experiencing an increase (e.g. high latitudes and large parts of the USA) and others experiencing a decrease (e.g. parts of West Africa, southern Europe and southernmost South America) (Bates et al., 2008). In regions where winter precipitation falls in the form of snow, there is more robust evidence that the timing of river flows has been significantly altered.

Projections indicate increases in runoff in high latitudes and the wet tropics, and decreases in mid-latitudes and some parts of the dry tropics. In some simulations, runoff is notably reduced in southern Europe and increased in south-east Asia and in high latitudes, where there is consistency among models in the direction of the change, although less so in terms of the magnitude of change (Bates et al., 2008).

Increasing frequency and intensity of extreme events[5]

Climate change leads to changes in the frequency, intensity, spatial extent, duration, and timing of extreme weather and climate events, and can result in unprecedented extreme weather and climate events (IPCC, 2012). Extreme events have an impact on water quantity and quality and pose risks to populations, physical assets and ecosystems. Water quality is expected to be negatively impacted by extreme events via increasing sediments, nutrients, pesticides and salt loads. The risk of contamination of water supplies due to increased microbial loads spurred by both extreme rainfall and droughts could increase disease outbreaks.

Changes in extremes can be linked to changes in the mean, variance or shape of probability distributions, or a combination of all of these. Not all climate extremes are "tail events".[6] Some climate extremes, such as **drought**, may be the accumulation of weather or climate events that are not considered extreme when considered independently. Many extreme events continue to be the result of natural climate variability. Records in most places are patchy, which can mask extreme events, such as mega droughts in West Africa. A recent study of the paleoclimate record in Lake Bosumtwi, Ghana suggests that the most recent centennial scale drought took place only 200 to 300 years ago. This event was much more severe and longer than the multidecadal drought of the 1970s that had widespread ecological, political, and socioeconomic impacts (Shanahan et al., 2009). Natural variability will continue to be an important factor in shaping future extremes in addition to the effect of anthropogenic changes in climate[7] (IPCC, 2012).

Observations indicate that some regions, in particular in southern Europe and West Africa, have experienced trends toward more intense and longer **droughts** since the 1950s (IPCC, 2012). However, in some regions (e.g. Central North America and North Western Australia) droughts have become less frequent, less intense, or shorter. Recent studies project an increase in the duration and intensity of droughts in some regions, including southern Europe and the Mediterranean region, Central Europe, Central North America, Central America and Mexico, Northeast Brazil, and Southern Africa. These shifts may contribute to positive feedback mechanisms in ecosystems. For example, two recent droughts in the Amazon demonstrate mechanisms by which remaining intact tropical forests of South America can shift from buffering the increase in atmospheric carbon dioxide to accelerating it (Lewis et al., 2011).

Figure 1.3 depicts projections of changes in the frequency and intensity of water resources drought across Europe by the 2070s. Severe droughts are projected to become less frequent and intense in some areas of Northern Europe and more frequent and intense in some areas of Southern Europe (IPCC, 2012). The figure also illustrates that while the two models reflect broadly similar trends, at the finest level of resolution, there are significant differences in the projections.

The number of **heavy precipitation** events has probably increased in many regions, but there are strong regional and sub-regional variations in these trends. North America exhibits the most consistent trends toward heavier precipitation. Projections indicate that it is likely that the frequency of heavy precipitation events will increase over many areas, especially in the high latitudes and tropical regions, and northern mid-latitudes in winter (IPCC, 2012).

As a result of increased frequency and intensity of rainfall events, **floods** are likely to become more frequent and potentially more severe in most regions around the world (Milly et al., 2002). It is difficult, however, to pinpoint the impact of climate change on observed changes in flooding, due to limited data and because of the importance of other factors that drive flood risk, such as changes in land use.

Impacts on freshwater ecosystems

Changes to the freshwater flow regime will be the most significant and pervasive of the impacts of climate change on freshwater ecosystems (Le Quesne et al., 2010). The integrity of flowing water systems depends largely on their natural dynamic character, with stream flow quantity and timing being the critical components (Poff et al., 1997). As such, changes in the timing of flows, as much as changes to total annual runoff, are likely to have the most significant consequences on freshwater ecosystems (Le Quesne et al., 2010). Although all ecosystems are threatened by climate change, freshwater ecosystems are especially vulnerable, with one of the highest numbers of threatened species (Ludwig and Monech, 2009).

Figure 1.3. **Projected change in indicators of water resources drought across Europe by the 2070s**

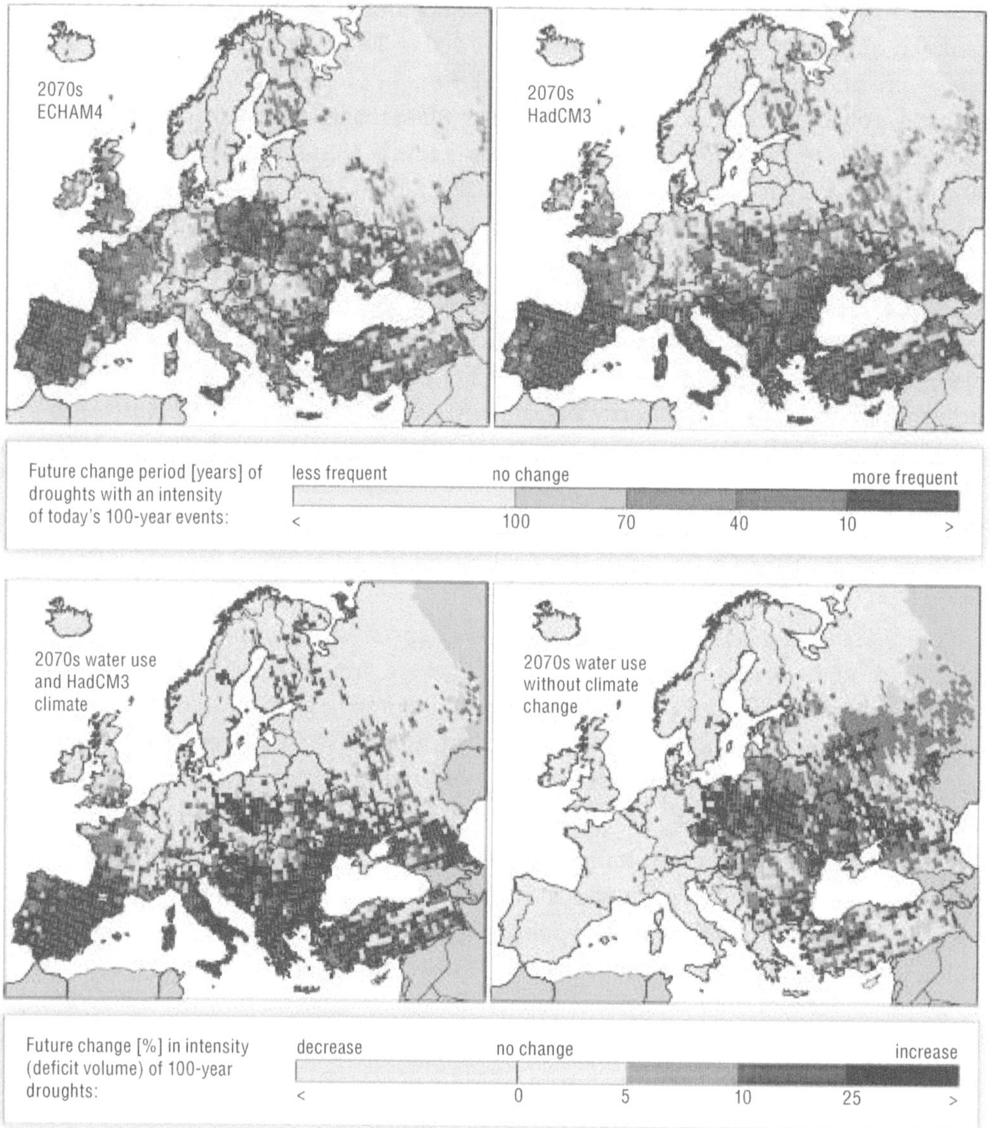

Top panel: projected changes in the return period of the 1961-90 100-year drought deficit volume for the 2070s, with change in river flows and withdrawals for two climate models, ECHAM4 and HadCM3. *Bottom panel:* projected changes in the intensity (deficit volume) of 100-year droughts with changing withdrawals for the 2070s, with climate change (*left*, with HadCM3 climate projections) and without climate change (*right*).
Source: Lehner, B. et al. (2006), "Estimating the Impact of Global Change on Flood and Drought Risks in Europe: A Continental, Integrated Analysis", *Climatic Change*, Vol. 75/3, Springer, pp. 273-299, *http://dx.doi.org/10.1007/s10584-006-6338-4*. Reprinted with kind permission from Springer Science + Business Media B.V.

Ecosystems respond to changes in hydrology in complex and often non-linear ways. Climate change impacts on ecosystems will occur through dramatic state shifts as "tipping points" are crossed as well as through gradual deterioration (Le Quesne et al., 2010). Evidence from recent studies by paleoecologists suggests that climate change may not simply result in mass migration of species, but instead, reshuffle into novel "no analog" ecosystems unknown today (Fox, 2007).

Rising sea level

Global mean sea level has been rising and there is high confidence that this has occurred at an increasing rate between the mid-19th and the mid-20th centuries (Bates et al., 2008). The spatial distribution of changes is uneven. For example, in the period 1993-2003, some regions experienced sea level rise up to several times the global mean, while sea level fell in other regions (Bates et al., 2008). In Hudson Bay, for instance, there is sea-level retreat, caused by the buoyancy of the land surface following the end of the last glacial period. The destruction of coastal wetlands or the modification of flow volume or speed can accelerate the rate of sea level rise. Changes in sea level tend to be very gradual (e.g. measured in millimetres annually), then may increase suddenly during storms, as weakened coastal defences give way.

In coastal areas, sea level rise will extend areas of salinisation of groundwater and estuaries, reducing the availability of coastal freshwater supplies. Sea-level rise also obstructs drainage in deltas.

Changes in glaciers, snow, ice and permafrost

The majority of glaciers and ice caps worldwide have shrunk, leading to significant declines in water storage in mountain glaciers and snow cover and very likely making a contribution to observed sea-level rise (Bates et al., 2008). In most regions, snow cover has decreased, especially during spring and summer. Warming has also shifted the timing and amplitude of runoff in glacier- and snowmelt-fed rivers and of ice-related phenomena in rivers and lakes. The degradation of permafrost and seasonally frozen ground negatively affects drainage systems. These changes have implications for reservoir storage, water supply, drainage systems, hydropower generation and flood management. The methane emissions from defrosting decayed plant matter also constitute a positive feedback mechanism for climate change, as methane is a very powerful greenhouse gas.

Climate change is expected to further reduce water supplies stored in glaciers and snow cover. Shifts in the flow regimes of rivers and streams of snow and glacier-fed basins (especially the timing of melt water) are projected, reducing water availability during warm and dry periods. Projections indicate widespread reductions in snow cover throughout this century, despite some projected increases at higher altitudes. Glaciers and ice caps are projected to shrink, with an eventual reduction in glacial melt water, with the possibility that some glaciers may disappear altogether. Glacial retreat may also result in glacial lake expansion, flooding from melting of glaciers and bursting of glacial lakes (Bates et al., 2008).

Increasing water temperature

Higher water temperatures are projected to aggravate water pollution by increasing pathogens, dissolved organic carbon, and thermal pollution. Algal blooms will occur more frequently in a warmer climate and higher temperatures will also increase microbial activity and bacterial and fungal populations (Ludwig and Monech, 2009). These effects compound with negative impacts on water quality from more frequent extreme events and lower flow conditions during some seasons, which can increase the concentration of pollutants. These changes are projected to exacerbate many forms of existing water pollution, with potentially detrimental effects on freshwater ecosystems, human health, and the operation of water systems. However, in some regions with increasing precipitation (e.g. northern Europe), water temperatures are declining, given the greater mass in water bodies.

Groundwater

Climate change will affect the depth of groundwater tables and the amount of groundwater available through changes in recharge rates. Both changes in annual rainfall amount and extremes will impact on groundwater recharge rates (Ludwig and Moench, 2009). Groundwater recharge may also be affected by increasing evapotranspiration. Although research on this topic has improved in recent years, much uncertainly remains as to how climate change will affect groundwater.

Impacts on evapotranspiration and soil moisture

Climate change has the potential to affect evapotranspiration and soil moisture. In some areas, the rate of evapotranspiration can be quite high, driving dry season scarcity. Yet, in general, there are very limited direct measurements of actual evapotranspiration, and hence, sparse information on observed trends (Bates et al., 2008). In principle, as temperatures rise, the capacity of the atmosphere to hold water increases, thus "potential evaporation" is projected to increase almost everywhere, yet other climate factors may exaggerate or offset this effect (IPCC, 2007).

Changes in evaporation and the amount and timing of precipitation will have an impact on soil moisture. However, data on changes in soil moisture are very sparse and the magnitude of projected changes is uncertain. Decreases are projected in some areas (e.g. sub-tropics, the Mediterranean, and high latitudes) while increases are projected in others (e.g. East Africa, central Asia) (Bates et al., 2008).

Concluding remarks

The scientific evidence of the range and complexity of climate change impacts on freshwater resources is compelling and growing. Given that water is an essential resource as well as a potential threat, climate change impacts on freshwater will affect not only water and flood management *per se*, but also a number of key policy domains (e.g. energy, agriculture, infrastructure, biodiversity, and health). Despite the ever-expanding scientific basis, reliable information about the nature, magnitude and timing of impacts on freshwater at the scale needed for practical, site-specific adaptation planning is lacking. Adaptation decisions need to accommodate significant uncertainty.

Effective and timely adaptation can lower the cost of climate change impacts by improving the management of water risks and enhancing resilience. In both the near and longer term, adaptation to climate change calls for flexible, dynamic, and future-oriented approaches that take into account climate variability on all timescales along with associated changes in human and natural systems. In the absence of accurate and precise climate predictions, a risk-based approach can explicitly accommodate a range of possible futures to inform adaptation decision making. It can also be used to identify priorities and options to manage water risks and enhance resilience at least cost to society. Chapter 2 explores such an approach, in the broader context of achieving water security.

Notes

1. "Water systems" refer to natural and man-made systems used to manage water resources and floods, and to provide water supply and sanitation. This report focuses on freshwater. Climate change impacts such as sea-level rise are only considered to the extent that they impact on coastal freshwater supplies, e.g. through the salinisation of groundwater.
2. As expressed in Milly et al. (2008), stationarity is the idea that natural systems fluctuate within an unchanging envelope of variability. It is a foundational concept that permeates training and practice in water-resource engineering.
3. This difference is often more significant than differences arising from the same model run with different emission scenarios.
4. The theoretical basis for this intensification is summarised in the Clausius-Clapyeron relation that implies that specific humidity would increase approximately exponentially with temperature.
5. This section is based largely on the IPCC (2012) *Special Report on Managing the Risks of Extreme Events and Disasters to Advance Climate Change Adaptation.*
6. "Tail events" are those with a very low probability in a typical probability distribution curve ("bell curve").
7. Extreme events are rare, which means there are few data available to make assessments regarding change in their frequency or intensity. The more rare the event, the more difficult it is to identify long-term changes.

References

Anagnostopoulos, D. et al. (2010), "A Comparison of Local and Aggregated Climate Model Outputs with Observed Data", *Hydrological Sciences Journal,* Vol. 55/70, Taylor and Francis, London, pp. 1094-1110, *http://dx.doi.org/10.1080/02626667.2010.513518.*

Bates, B., Z. Kundzewicz, S. Wu and J. Palutikof (eds.) (2008), "Climate Change and Water", *Technical Paper VI,* Intergovernmental Panel on Climate Change (IPCC) Secretariat, Geneva.

Brown, C. (2010), "The End of Reliability", *Journal of Water Resources Planning and Management*, Vol. 136, American Society of Civil Engineers, pp. 143-145.

Brown, C. and R.L. Wilby (2012), "An Alternative Approach to Assessing Climate Risks", *Eos, Transactions,* Vol. 93/41, American Geophysical Union, Washington, DC, pp. 401-402, *http://dx.doi.org/10.1029/2012EO410001.*

Dessai, S. et al. (2009), "Do We Need Better Predictions to Adapt to a Changing Climate?", *Eos, Transactions*, Vol. 90/13, American Geophysical Union, Washington, DC, pp. 111-112.

Fox, D. (2007), "Back to the No-analog Future?", *Science*, Vol. 316, American Association for the Advancement of Science, pp. 823-825.

Huntington, T. (2006), "Evidence for Intensification of the Global Water Cycle: Review and Synthesis", *Journal of Hydrology*, Vol. 319, Elsevier, pp. 83-95. *http://dx.doi.org/10.1016/j.jhydrol.2005.07.003.*

Intergovernmental Panel for Climate Change (IPCC) (2012), "Managing the Risks of Extreme Events and Disasters to Advance Climate Change Adaptation", *A Special Report of Working Groups I and II of the Intergovernmental Panel on Climate Change*, Cambridge University Press, Cambridge, UK and New York, NY, USA.

IPCC (2007), "Climate Change 2007: Impacts, Adaptation and Vulnerability", *Working Group II Contribution to the Fourth Assessment Report of the Intergovernmental Panel on Climate Change*, Cambridge University Press, Cambridge.

Kundzewicz, Z.W. et al. (2008), "The Implications of Projected Climate Change for Freshwater Resources and their Management", *Hydrological Sciences Journal*, Vol. 53/1, Taylor and Francis, London, pp. 3-10.

Kundzewicz, Z.W. and L. Somlyódy (1997), "Climatic Change Impact on Water Resources in a Systems Perspective", *Water Resources Management*, Vol. 11, Kluwer Academic Publishers, the Netherlands, pp. 407-435.

Kundzewicz, Z.W. and E.Z. Stakhiv (2010), "Are Climate Models 'Ready for Prime Time' in Water Resources Management Applications, or is More Research Needed?", *Hydrological Sciences Journal*, Vol. 55/7, Taylor and Francis, London, pp. 1085-1089. *http://dx.doi.org/10.1080/02626667.2010.513211.*

Le Quesne, T. et al. (2010), "Flowing Forward: Freshwater Ecosystem Adaptation to Climate Change in Water Resources Management and Biodiversity Conservation", *Water Working Note*, No. 28, World Bank Group, Washington, DC.

Lehner, B. et al. (2006), "Estimating the Impact of Global Change on Flood and Drought Risks in Europe: A Continental, Integrated Analysis", *Climatic Change*, Vol. 75/3, Springer, pp. 273-299, *http://dx.doi.org/10.1007/s10584-006-6338-4*.

Lewis, S.L. et al. (2011), "The 2010 Amazon Drought", *Science*, Vol. 331/6017, American Association for the Advancement of Science, p. 554, *http://dx.doi.org/10.1126/science.1200807*.

Ludwig, F. and M. Moench (2009), "The Impacts of Climate Change on Water", *in* F. Ludwig, P. Kabat, H. van Schaik and M. van der Valk (eds.), *Climate Change Adaptation in the Water Sector*, Earthscan, London and Washington, DC., pp. 35-50.

Matthews, J., A.J. Wickel and S. Freeman (2011), "Converging Currents in Climate-Relevant Conservation: Water, Infrastructure, and Institutions", *PLoS Biology*, Vol. 9/9, PLOS, *http://dx.doi.org/10.1371/journal.pbio.1001159*.

Milly, P.C.D. et al. (2008), "Stationarity is Dead: Whither Water Management", *Science*, Vol. 319, American Association for the Advancement of Science, pp. 573-574, *http://dx.doi.org/10.1126/science.1151915*.

Milly, P.C.D. et al. (2002), "Increasing Risk of Great Floods in a Change Climate", *Nature*, Vol. 415, Nature Publishing Group, pp. 514-517.

OECD (2012), *OECD Environmental Outlook to 2050: The Consequences of Inaction*, OECD Publishing, *http://dx.doi.org/10.1787/9789264122246-en*.

Poff, N.L. et al. (1997), "The Natural Flow Regime", *BioScience*, Vol. 47/11, American Institute of Biological Sciences, pp. 769-784.

Rodríguez-Iturbe, I. and J.B. Valdés (2011), "Uncertainty and Climate Variability in the Design and Operation of Water Resources Projects: Examples and Case Studies", *HEF Technical Report 2*, The World Bank, Washington, DC.

Shanahan, T.M. et al. (2009), "Atlantic Forcing of Persistent Drought in West Africa", *Science*, Vol. 324/5925, American Association for the Advancement of Science, pp. 377-380, *http://dx.doi.org/10.1126/science.1166352*.

United Nations Framework Convention on Climate Change (UNFCCC) (2011), "Water and Cclimate Change Impacts and Adaptation Strategies", *Technical Paper*, 22 November, FCCC/TP/2011/5, *http://unfccc.int/resource/docs/2011/tp/05.pdf*.

Wickel, A.J., and J.H. Matthews (2012), Adapted from "Vulnerability to What Change?", presented at the UNFCCC Technical Workshop on Water, Climate Change Impacts and Adaptation Strategies, Mexico City, Mexico, 18-20 July, World Wildlife Fund and Conservation International, *http://unfccc.int/adaptation/workshops_meetings/nairobi_work_programme/items/6955*.php (accessed 12 December 2012).

Wilby, R.L. (2010), "Evaluating Climate Model Outputs for Hydrological Applications", *Hydrological Sciences Journal*, Vol. 55/7, Taylor and Francis, London, pp. 1090-1093, *http://dx.doi.org/10.1080/02626667.2010.513212*.

Wilby, R.L. and S. Dessai (2010), "Robust Adaptation to Climate Change", *Weather*, Vol. 65/7, Royal Meteorological Society, Reading, pp. 180-185, *http://dx.doi.org/10.1002/wea.543*.

Chapter 2

A risk-based approach to adapting water systems to climate change

This chapter sets out a risk-based approach for adapting water systems to climate change as a key component of improving water security over the long-term. It begins by setting out a rationale for a risk-based approach to inform policy decisions in the absence of reliable climate predictions. It then provides a framework to guide government action to "know", "target", and "manage" water risks and highlights the importance of considering "risk-risk trade-offs". Finally, it proposes policy guidelines for addressing water risks in the context of a changing climate in order to improve the timeliness, efficiency, and equity of adaptation decisions.

Key messages

- A **risk-based approach** provides a flexible, dynamic and future-oriented approach to managing water risks in the absence of accurate and precise climate predictions. Adaptation should not be undertaken in an isolated way, by focussing only on climate as a risk driver to the exclusion of other, often more dominant, drivers of water risks. At the same time, failure to adapt to climate change can jeopardise water security over the long-term and make it more costly to improve.
- Adaptation is **not about maintaining the status quo at all costs.** It is about adjusting to new circumstances in order to reduce adverse effects, minimise costs and seize any potential opportunities.
- Although the cost and benefits of adapting water systems may be difficult to assess with precision, the **cost of not adapting can be high**, especially in a climate-sensitive policy domain such as water, with long-lived, capital-intensive infrastructure and high sunk costs.
- Efforts to reduce a given water risk (e.g. shortage) for one group of users, such as farmers, may increase other water risks (e.g. undermining the resilience of freshwater systems) affecting other users, such as fish. Weighing these "**risk-risk trade-offs**" can help to identify "win-win" strategies and to reduce inefficiencies and inequities in managing water risks.
- In a risk-based approach, there is a need to *"know", "target" and "manage"* water risks.
  - To "**know**" the risk requires providing information to reduce information asymmetries and form the basis for making effective and informed risk management decisions.
  - To **"target"** the risk requires facilitating stakeholders' agreement on the *acceptability* and *tolerability* of a given risk, relying on both evidence-based and values-based judgements.
  - To "**manage**" the risk requires clarity in terms of risk sharing arrangements between public and private actors to ensure that the risk is managed by actors able to do so most efficiently. Governments need to consider the full range of risk management strategies (avoid, reduce, transfer, bear) and tap the broad range of policy instruments to facilitate timely and efficient adaptation.

Why risk? Rationale for a risk-based approach[1]

Freshwater systems are inherently variable. The availability of water resources varies in time (seasonal, inter- and intra-decadal variability) and space as a result of the natural water cycle. Climate change reinforces and exacerbates these features by accelerating the water cycle (Huntington, 2006) and increasing uncertainty over the distribution of water resources in time and space (Milly et al., 2008).

Water quality also depends on a range of constantly fluctuating parameters, including the nature and quantity of pollutants, extent of dilution in water bodies, and the likelihood of harmful impacts. Extreme events, such as floods and droughts, have significant destructive potential, but the location of their occurrence, timing, and magnitude can only be estimated.

In the absence of accurate and precise climate predictions, a risk-based approach can be used to explicitly accommodate a range of possible futures. It can be used to identify the range of options to manage water risks and help to prioritise responses. Risk analysis encourages policy makers to consider a whole range of possible future conditions, from the commonplace to the extremely unlikely. Thinking in terms of risk has immediate relevance for decision making, where alternative actions or activities (including the choice to do nothing) are weighed (Hall, 2012). Risk concepts are increasingly used to understand and to inform behavioural change and public and private sector investments to enhance water security (Loucks, 2011). Indeed, risk approaches are widely discussed in the adaptation literature and are gathering increasing attention from policy makers (see examples from OECD countries in Chapter 3).

Although climate change is a significant driver of water risks, there are a number of reasons why adaptation to climate change should be considered in the broader context of water security and not as an isolated issue. First, climate change is just one of many drivers of water risks, along with socio-economic trends, natural phenomena, and poor water governance. In most cases, it is not the dominant risk driver, nor expected to be so for the near to medium term. Second, natural variability in the climate confounds attempts to attribute specific events to long-term climate change and to neatly differentiate between the effects of natural climate variability and long-term climate change. Finally, many adaptation policy responses for water systems are synonymous with good water policy and management, more generally. Thus, what does and does not fall within the purview of "adaptation" is often ambiguous, much more than in the case of climate change mitigation.

As a result, there is a general recognition that climate change adaptation should be "mainstreamed" into existing water resources management rather than based on a "climate-driven" approach.[2] Indeed, examples of concrete actions to adapt water systems specifically and solely to a changing climate are rare. As a result, an excessive focus on attributing impacts on water systems due to climate change in order to determine adaptation responses may, in fact, inhibit sensible approaches.[3] However, a focus of mainstreaming should not diminish the importance of climate change adaptation, either. A failure to systematically integrate scientific evidence of climate change into water management and to prepare for a more uncertain future is likely to become increasingly costly as time goes on. Indeed, climate change adaptation has become a requisite to realising water security over the long-term.

Achieving water security requires maintaining an *acceptable* level of water risks – in terms of water shortage, excess, pollution, and freshwater system resilience – for society and the environment, today and in the future, through the effective and efficient application of water and water-related policies (OECD, *forthcoming*).[4] The main thrust of a risk-based approach to water security is to secure benefits for society and the environment in a way that maximises expected social welfare. While the risk-based approach is not the only approach to water security, it a sensible one in that it can help to prioritise improvements at the margin that maximise net expected benefits. Water has diminishing marginal value. Once water demand is satisfied, additional units of water have comparatively little value. Once the quality of water is sufficiently fit for use, additional increments of quality improvement are of comparatively little value. Once an acceptable level of flood protection is attained, there is little additional benefit to building higher flood barriers. Once water risks are tolerably low, then water is hardly a concern (Hall, 2012).

Governments cannot guarantee "zero risk" and reducing or avoiding risk generally comes at a cost. This cost may be in economic (e.g. the cost of building infrastructure to increase water storage or to augment flood protection; or the opportunity cost of foregone spatial development in flood plains), social (e.g. closing water allocations in a basin to cap demand), and/ or environmental terms (e.g. deterioration of freshwater systems due to excessive diversions).

In the same vein, adapting to climate change is not equivalent to maintaining the status quo at all costs. While climate change adaptation can lower the overall costs of the impacts of climate change, many adaptation measures also come at a cost. Depending on the water risks faced, incremental risk reduction may be disproportionally costly. By identifying the level of acceptability and tolerability of water risks, a risk-based approach fosters targeted and proportional policy responses, and thus cost effectiveness. It also helps to clarify risk-sharing arrangements between public and private actors. Water management is an inherently risky business, where although it is in the *interest* of most stakeholders to minimise the water risks, it is not the *responsibility* of most stakeholders.

Thinking risk: A framework

A framework laying out the key steps involved in risk assessment and management provides a guide to a risk-based approach to water security (Figure 2.1). The framework is comprised of four main components, described in the following section.

Which water risks? Identifying water risks and risk drivers

Risks can be defined in many different ways and the domain of water policy is no exception. Risks arise not only from physical phenomena but also from the social context. Societies vary in the ways in which they select which problems are identified as risks, and which risks require attention and response. Different actors within societies define risks in different ways. Too much or too little water can be considered a risk, depending on the level of water required. For instance, a certain level of annual flooding might be necessary for a productive agricultural cycle. Levels of inundation that either fall below or exceed the desired range would be considered a risk. In Ancient Egypt, a moderate annual inundation was a vital part of the agricultural cycle. A lighter inundation would cause famine, while excessive flooding would cause destruction of infrastructure built on the flood plain.

Figure 2.1. **A risk-based framework for water security**

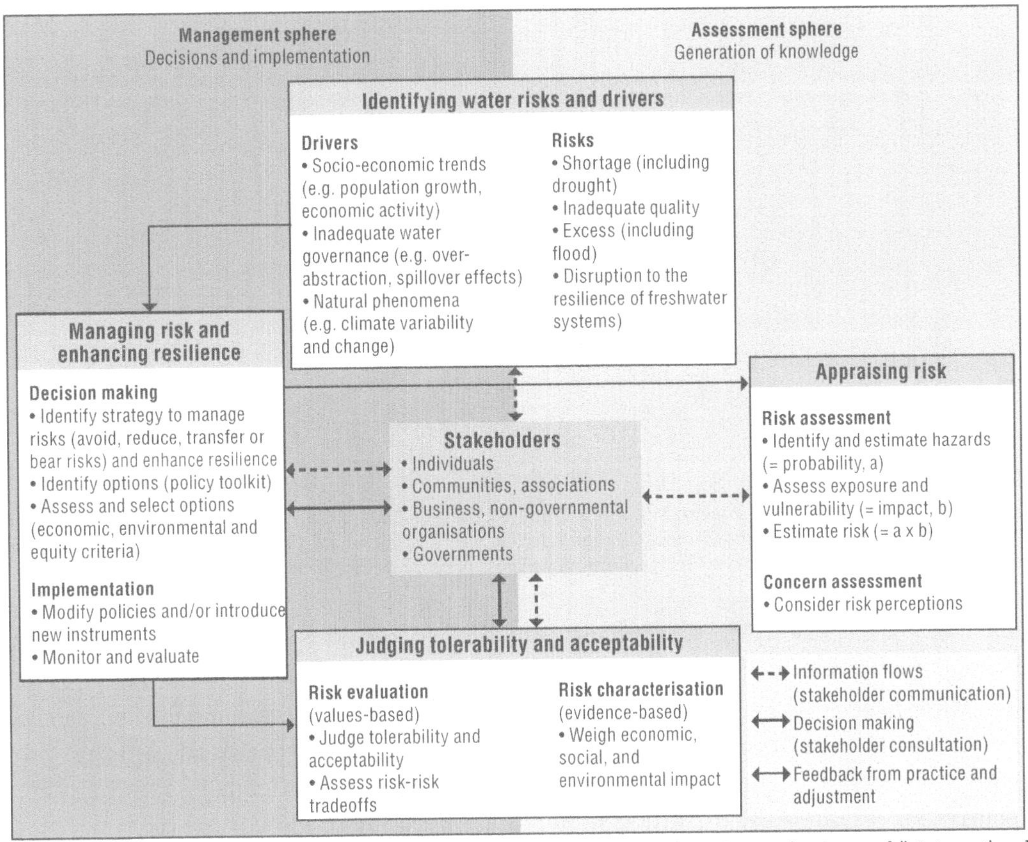

Source: Adapted from O. Renn and P. Graham (2006), "Risk Governance: Towards an Integrative Approach", International Risk Governance Council (IRGC), *White Paper No. 1*, *www.irgc.org/IMG/pdf/IRGC_WP_No_1_Risk_Governance__reprinted_version_.pdf* (accessed 4 December 2012).

This framework sets out four water risks. Achieving water security means maintaining *acceptable* levels of all of these risks:

- ***Risk of shortage*** *(including droughts)*: lack of sufficient water to meet demand (in both the short- and long-run) for beneficial uses by all water users (households, businesses and the environment).
- ***Risk of inadequate quality:*** lack of water of suitable quality for a particular purpose.
- ***Risk of excess*** *(including floods)*: overflow of the normal confines of a water system (natural or built), or the destructive accumulation of water over areas that are not normally submerged.
- ***Risk of undermining the resilience of freshwater systems:*** exceeding the coping capacity of the surface and groundwater bodies and their interactions (the "system"), possibly crossing tipping points, and causing irreversible damage or system collapse.

All four risks should be considered at the same time as they can impact on each other, given the interconnected nature of water resources. Indeed, these risks are interrelated. For instance, the risks of shortage, inadequate quality and excess may all increase the risk of undermining the resilience of freshwater systems.

Water risks arise not only from the potential consequences and likelihood of *hazards*, such as extreme weather events, but also from drivers of *exposure* and *vulnerability* to risk (Figure 2.2). Poor water governance is considered the major driver of water risks (Grafton

Figure 2.2. **Risk at the intersection of hazard, exposure and vulnerability**

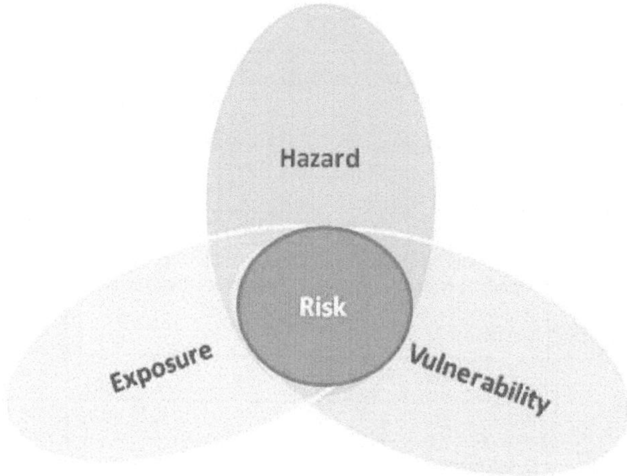

Source: Adapted from Intergovernmental Panel for Climate Change (IPCC) (2012), "Managing the Risks of Extreme Events and Disasters to Advance Climate Change Adaptation", *A Special Report of Working Groups I and II* of the Intergovernmental Panel on Climate Change, Cambridge University Press, Cambridge, UK and New York, NY, USA.

et al., 2013). For example, it may lead to lack of adequate water infrastructures and technology, due to neglect, insufficient financing, and/ or poor asset management.

Water risks often result from spill over effects from other policy domains. By creating incentives to meet their own objectives, sectoral (e.g. agricultural, energy) and environmental (e.g. climate, biodiversity) policies have significant effects on water security. For example, by distorting production and trade of agricultural commodities, agricultural policy distorts the domestic demand for water.

The degradation of ecosystems or "natural infrastructure" can increase the vulnerability of populations, ecosystems, and physical assets to water risks. For example, wetlands or mangroves can perform vital ecosystem services to reduce flood risk. The degradation or destruction of this "natural infrastructure" increases the vulnerability of the populations or physical assets exposed to flooding, thus increasing risk. Natural infrastructure solutions can be used to "retrofit increased risk reduction" into highly managed river basins (Coates and Smith, 2012).

From an efficiency perspective, the management of water risks should focus on events with the most impact. There has been implicit assumption that this means focusing on extreme ("tail") events with low probability and high impact, such as extreme floods. In the case of climate change, these are likely to be the most severe, costly and difficult to predict. But the long-term consequences of "normal" (high probability, low initial impact), but recurrent or chronic, threats to water security, such as competition for water resources among various users or pollution, deserves much greater attention from a risk perspective. These concealed or dormant risks develop slowly and are thus often considered as "invisible", with their main impacts emerging only in the long-term. Yet, there are subtle signs that may signal risk triggers that can cause major shifts. These signals include slower recovery from small disturbances (known as "critical slowing down").[5] Addressing these risks enhances the resilience of water systems to better respond to long-term pressures, including climate change.

"Know" the risk: Appraising water risks

Building an adequate information base to inform decisions about water risks requires bringing together two components: a scientific risk assessment and an understanding of risk perceptions by stakeholders. Water management decisions typically rely predominately on science and technology (engineering approaches), underestimating the importance of social, economic and political drivers of risk (Rees, 2002). This is particularly important to keep in mind for climate change adaptation, where there is often a temptation to focus mainly on climate change *hazards*, shifting attention away from other (often more dominant) drivers of water risks, including factors affecting vulnerability and exposure to those risks. This narrow focus can significantly limit the overall effectiveness and efficiency of adaptation responses. While scientific and technical inputs are important elements in a risk-based approach, economic, social and cultural dimensions also need to be considered to ensure that policy responses are proportionate, economically efficient, and equitable.

Although good physical science and technical expertise are a prerequisite for sound risk assessment, they alone cannot be the main basis for decision-making (Rees, 2002). The understanding of risk perceptions, via a *concern assessment* is a fundamental (and often overlooked) step in the risk appraisal process. It is a key element in seeking to assign clear roles and responsibilities for managing risks. For example, farmers' perception of drought risk has an important influence on their decisions affecting their vulnerability to drought and their risk management strategies. Concern assessment can also help to inform public debates addressing issues of "contested values" that are often at the heart of conflicts over water (e.g. determining sustainable levels of use and/or pollution).

The conventional risk assessment process, which typically places scientific assessment at the starting point for analysis, can also be inverted to place the human context, knowledge, needs and preferences as the first stage of appraisal. Such "inverted" risk appraisal models could help to ensure that risk management becomes more demand responsive and more inclusive. However, incorporating risk perceptions into the risk appraisal process raises a major issue in terms of how to deal with subjective beliefs.

A formal risk appraisal process for water risks can be data-intensive and costly in terms of time and resources. Often, significant scientific capacity is needed. In cases where significant populations or assets are at risk (e.g. densely populated urban areas in flood plains) and the cost of risk reduction is significant (e.g. structural flood protection), a formal and comprehensive risk appraisal is justified. In cases where the stakes are lower, a less formal, but still informative, qualitative assessment or rapid risk assessment may be sufficient. The depth and extent of the appraisal undertaken should be proportional to the magnitude of the risk.

Decision making under uncertainty

In appraising water risks, there will be many cases where the scientific assessment will be limited by sparse data, knowledge gaps and other sources of uncertainty (Box 2.1). For certain risks, science will never know all of the hydrological facts at stake. This is particularly acute in the case of climate change where, although the scientific evidence base is improving, pervasive uncertainty means that the "predict then act" approach to decision-making does not apply. As discussed in Chapter 1, typical assessments of the impacts of climate change on freshwater are of limited use when it comes to making practical, on-site decisions about adaptation.

Box 2.1. **Risk and uncertainty**

A common distinction between risk and uncertainty derives from Knight's (1921) observation that risk is uncertainty that can be reliably measured. Thus, risk describes the likelihood and consequence of an uncertain event of which the probability of occurrence can be reliably estimated. Uncertainty describes situations where the probability of occurrence is not known and perhaps cannot be known.

Uncertainty can arise from a lack of information or disagreement about what is known or even knowable. Sources of uncertainty include errors in data, lack of knowledge regarding the interaction of key variables, or uncertain future projections of human or natural systems.

When it comes to the process of risk assessment, the systematic characterisation of uncertainties is a key challenge. A key step is to identify the sources of uncertainty and to be explicit about the degree of confidence experts have in the knowledge base. Uncertainty can be characterised quantitatively, for example, by a range of values calculated by various models assessed with various confidence intervals. It can also be expressed qualitatively, by reflecting expert judgement. In some cases, uncertainty can be reduced by more scientific research, while in other cases uncertainty will persist regardless of advances in scientific knowledge. Advances in scientific understanding may also generate new uncertainties, as a better understanding of complex systems is gained.

Source: F.H. Knight (1921), *Risk, Uncertainty, and Profit*, Hart, Schaffner and Marx, Houghton Mifflin Company, Boston; Adapted from O. Renn and P. Graham (2006), "Risk Governance: Towards an Integrative Approach", International Risk Governance Council (IRGC), *White Paper*, No. 1, *www.irgc.org/IMG/pdf/IRGC_WP_No_1_Risk_Governance__reprinted_version_.pdf* (accessed 4 December 2012); Intergovernmental Panel for Climate Change (IPCC) (2012), "Managing the Risks of Extreme Events and Disasters to Advance Climate Change Adaptation", *A Special Report of Working Groups I and II* of the Intergovernmental Panel on Climate Change, Cambridge University Press, Cambridge, UK and New York, NY, USA.

In cases of pervasive uncertainty, a number of alternative decision-making approaches can be used to improve the robustness of decisions, so that they work well under a range of possible futures. Scenario-led strategies can incorporate expert knowledge and stakeholder views to consider multiple possible futures. While planning for *all* possible futures is potentially very costly and planning for an *average*[6] possible future is potentially very wrong, the sensible use of scenario approaches can help to avoid over- (and under-) investment. Scaling investments in infrastructures incrementally over time can allow for adjustment as time goes on and more information is gained. Hallegate et al. (2012) provide a recent review of methodologies to inform investment under deep uncertainty in the context of climate change. Methods reviewed include robust decision-making, real options approaches, and cost-benefit analysis under uncertainty. Antón et al. (2012) summarises several approaches to evaluating decision-making under uncertainty (Box 2.2).

Brown and Wilby (2012) set out a strong case for the utility of alternative approaches to assessing climate risks based on sensitivity analysis (e.g. "decision scaling"). Recognising the limitations of using GCM projections to inform adaption decisions for water, these approaches use the projections to inform, rather than to drive the analysis. They could be described as "bottom-up meets top-down," as they focus first on the issues of concern and then on how climate information might add value to the analysis. These methods follow three basic steps: 1) identify the problem, including defining objectives and performance measures, 2) use a stress test to identify the hazard and evaluate the performance of the system under a wide range of non-climatic and climate variability and change, and 3) evaluate the risk using climate information, including climate model projections. While

Box 2.2. Approaches to evaluating decision-making under uncertainty

Bayesian "probabilistic" approach

Uncertainty across scenarios can be handled through a standard Bayesian probabilistic approach. This approach consists of assigning probabilities to each scenario and obtaining a combined distribution of outcomes that accounts for the probability of each different scenario to occur. Decision-making can be based on standard expected utility theory with or without government risk aversion.

"Satisficing" criterion

The "satisficing" principle is based on the idea of ensuring a reasonably good outcome. Since it will be difficult for a single policy instrument to be optimal across all possible states of the world reflected in different scenarios, an analysis can be carried out to see if there is an instrument that performs "well enough" in all situations under consideration. This principle was introduced by Simon (1956) to describe behavior in situations of bounded rationality and incomplete information. It is plausible that there is no instrument that performs "well enough" across all scenarios. In this case, the criterion helps to show the policy maker which scenarios are the most disregarded under each choice.

MaxiMin criterion

MaxiMin criterion focusses on avoiding worst-case outcomes in an adverse state of the world. In other words, maximising the minimum outcome (von Neumann, 1928). The principle is to take the worst-case scenario for any given instrument and choose the instrument that has the highest value for cost-effectiveness indicator in its worst-case scenario. This criterion is very conservative, representing high uncertainty aversion, and has the advantage of choosing a single instrument across all scenarios.

Source: Adapted from Antón, J. et al. (2012), "A Comparative Study of Risk Management in Agriculture under Climate Change", *OECD Food, Agriculture and Fisheries Papers*, No. 58, OECD Publishing, *http://dx.doi.org/10.1787/5k94d6fx5bd8-en*.

these methods cannot reduce the uncertainty associated with climate change, they do attempt to clarify the effect of the uncertainty on decisions.

These various alternative approaches can be used to inform decisions and to evaluate policy options under pervasive uncertainty. They can be used to prioritise policies and investments for adaptation. The suitability of any particular approach will depend on the nature of water risk being addressed, the amount and quality of evidence available (e.g. climate science, economic analysis) and the risk aversion of the decision makers. A practical example is the UK Climate Change Risk Assessment (CCRA) (see Box 3.1), which uses the best scientific evidence available alongside well-established risk-based decision approaches to assess risks and decide how to respond.

"Target the risk": Judging tolerability and acceptability

Determining the *acceptability* and *tolerability* of a given water risk is one of the most challenging and controversial aspects of a risk-based approach (Klinke and Renn, 2012). This process relies on both evidence-based and values-based judgements. A risk is considered *acceptable* if the likelihood of exceeding a given risk threshold (e.g. health standard, tipping point of a freshwater system) is low and the impact of exceeding that threshold is low. In such cases, there is no pressure to reduce acceptable risks further, unless more cost-effective measures become available. In contrast, cost-effective measures are required to reduce *tolerable* risks to an acceptable level. It may also be tolerable for a risk to exceed an acceptable threshold, provided it is temporary and reversible. Due to their very high probability and/or high damage potential, *intolerable* risks are deemed unacceptable. Urgent action is needed to reduce them to an acceptable level. The acceptability and tolerability judgement process enables policymakers to prioritise risk management decisions when risks exceed acceptable levels (OECD, 2009).

Complex decisions regarding the acceptable or tolerable level of risk to freshwater systems are routinely faced in decisions about water risks, including determining minimum environmental flows or setting flood safety standards. Whether implicit or explicit, the judgement regarding the acceptability and tolerability of a given risk strongly influences the risk management strategy, the response adopted, the role of public policy, and the current and future cost of risk management. Economic analysis has an important role to play for the evidence-based judgement, including weighing potential risk-risk tradeoffs. The tolerability and acceptability judgement process varies from country to country in terms of how it is undertaken, by whom, and the extent to which key stakeholder groups are involved.

Fish or farmers? Weighing risk-risk tradeoffs

Climate change adaptation and achieving water security often entails weighing risk-risk tradeoffs, or "risk externalities". Efforts to manage a given risk for a given population may (inadvertently or not) increase other risks. For example, reducing flood risk for one community can increase flood risk for a neighbouring community. Reducing the risk of shortage for consumptive users can increase the risk of undermining the resilience of freshwater systems. The trade-off between risk of shortage and risk of flood can be a dilemma in the operation of dams. Risk-risk trade-off analysis can help policy makers recognise trade-offs that may result from a policy intervention (or lack thereof), weigh the comparative importance of managing interrelated risks when difficult choices are required, and analyse the possibility of overall risk reduction through "win-win" responses (Graham and Wiener, 1995).

The point of thinking about risk-risk trade-offs is to help identify strategies that minimise the negative externalities of risk management. When considered solely on their own, strategies for managing a given water risk can *appear* optimal, even if they entail significant externalities. However, taking into account the consequences on other water risks, the trade-offs becomes evident and the payoffs change. A simple example of a risk-risk trade-off can be illustrated graphically (Figure 2.3). This example illustrates the potential trade-off between farmers (e.g. the risk of shortage) and fish (e.g. the risk to undermining the resilience of freshwater systems).

In this example, increased diversions to reduce the risk of shortage for farmers can increase the risk of disrupting the resilience of freshwater systems. However, improving efficiency of irrigators' use can reduce the risk of shortage without increasing diversions and hence, have a neutral (or potentially positive) effect on the risk to freshwater systems.

Another way to visualise risk-risk trade-offs is by aggregating all of the possible risk-risk trade-offs for a given pair of risks, using the maximally effective available interventions. This can be illustrated by a "risk protection frontier" (Figure 2.4, left panel). The shape of the risk protection frontier depends on the relationship between risks. This will be distinct for any given set of risks and contextual factors, for example, the physical characteristics of water

Figure 2.3. **Illustration of a potential trade-off between fish and farmers**

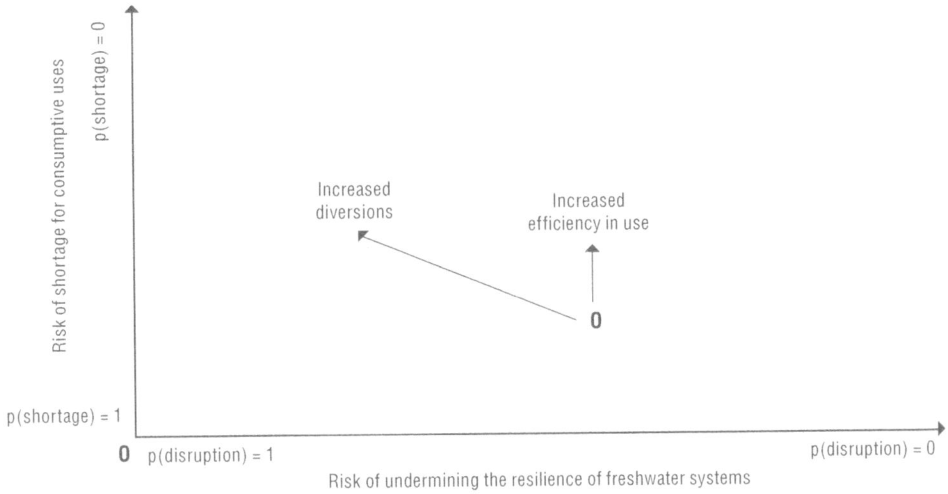

bodies as well as the technical possibilities employed to manage risk. In this illustration, the risk protection frontier is concave, reflecting an assumption that there are diminishing marginal returns to protection against each risk. In other words, this means that as greater levels of risk reduction are achieved for one risk, a greater increase in the other risk must be tolerated. The example in Figure 2.4 shows a trade-off in allocating water either towards increased diversions for consumptive uses or towards ensuring higher minimum environmental flows.

Figure 2.4. **Risk protection frontier for allocating water between fish and farmers**

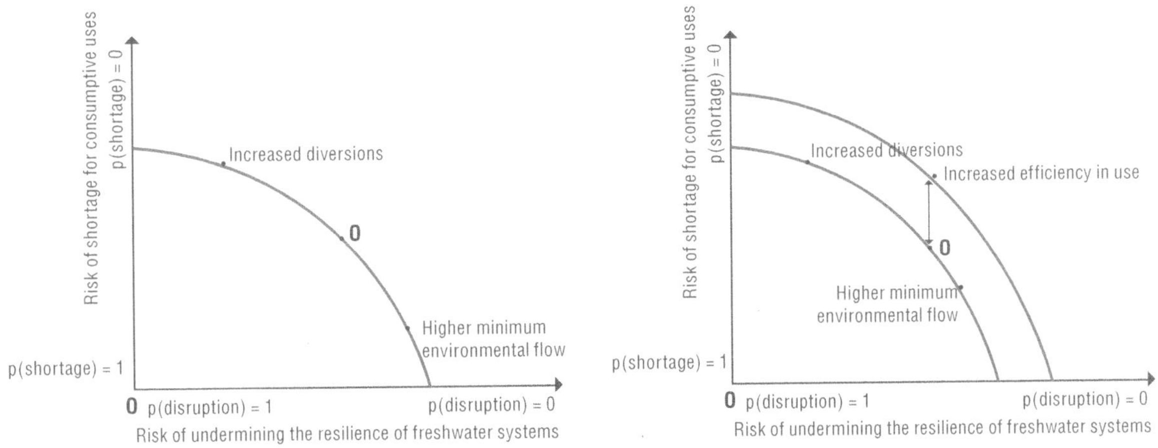

The risk protection frontier is not fixed. A risk management strategy that successfully achieves *overall* risk reduction for at least one or both risks, that does not require a trade-off between risks, can be illustrated by an outward shift (right panel, Figure 2.4), resulting, for example, from increased efficiency in use.

"Manage" the risk and enhance resilience

Decisions about managing water risks and enhancing resilience should be informed by all of the previous steps of the risk governance process. The strategy to manage risk may be to avoid, to reduce, to transfer or to bear the risk. This can be done by altering risk drivers, by limiting exposure or enhancing the resilience of the community, physical assets, and the environment by making them less vulnerable to potential harm. There are many possible adaptation strategies that can be used to manage water risks (illustrated in Table 2.1).

Table 2.1. **Possible adaptation strategies for water systems**

Adaptation strategy	Examples
Enhancing supply	Increased storage, rainwater harvesting, water re-use, recycling, desalinisation, reducing leakage in supply systems, groundwater recharge, extracting groundwater
Managing demand	Water pricing, smart water systems
Facilitating allocation of water resources	Water markets, administrative water allocation, connecting regional supply systems and facilitating inter-basin transfers
Maintaining or improving quality	Water quality standards, discharge limits, wastewater charges, ecosystems-based adaptation strategies (e.g. installation of riparian barriers, restoring/ protecting wetlands), improved water treatment plants
Improving flood protection	Land-use management; built or natural infrastructures (e.g. levees, wetlands management)
Improving information	Research and data collection and dissemination, climate change impact, vulnerability and risk assessments
Coping with uncertainty	Cyclical and iterative water management planning, scalable and potentially adaptable infrastructures (e.g. wetlands and floodplains, adaptive land use, a system of small dams versus one larger dam)

Adaptation can minimise the total cost of climate change (e.g. the cost of adaptation plus residual climate change damage). From an economic perspective, it is possible to have too much or too little adaptation, as well as mal-adaptation. In the context of risk and uncertainty, decisions are informed by weighing *expected* costs and benefits – that is, the probability-weighted mean over the range of possible outcomes. In the case of climate change adaptation, the cost of actions is more likely to be known and incurred in the near term, while many of the benefits (avoided climate impacts) will accrue far into the future and will not be known with certainty. Assessments of the costs and benefits of adaptation measures for water systems are important to inform decisions about prioritising adaptation responses and to improve their cost-effectiveness. Yet, the evidence base estimating the costs and benefits of adapting water systems remains sparse (Box 2.3).

The timing of adaptation actions will have an important impact on the overall costs and benefits of adaptation. In principle, decisions makers considering the timing of adaptation actions will compare the present value of adaptation today with the present value of adaptation at a later stage. The decision to act earlier or later will essentially depend on three factors: *i)* the difference in adaptation costs over time; *ii)* the short-term benefits of adaptation; and *iii)* the long-term effects of early adaptation (Agrawala and Fankhauser, 2008).

While the effect of discounting would generally favour delaying adaptation actions, early action may be cheaper for certain types of measures. They include making adjustments to long-lived infrastructure investments thus avoiding the cost and inconvenience of expensive retrofits. Early adaptation actions may also be preferred if they bring immediate and/or lasting benefits or avoid crossing irreversible tipping points (Agrawala and Fankhauser, 2008).

In practice, the timing and nature of adaptive actions taken are influenced by a range of factors. Hanemann (2008) points to a number of these: most adaptation is local; decisions involve multiple actors; the allocation of costs and benefits of adaptation and the pace of decision making are mediated by institutions; the facts of climate change and potential adaptive responses are not known with certainty, nor are they always agreed to

Box 2.3. **Expected costs and benefits of adapting water systems to climate change**

The cost of adapting to climate change will likely add to the already substantial financing gap for water systems in OECD countries. Greater variability in weather is likely to pose a bigger challenge for adapting water systems than shifts in averages, and will likely be a key cost driver for adapting water infrastructures.

Economic assessments of adapting water systems to climate change remain sparse. There are only a limited number of national, river basin or sub-national studies on adapting water supply. These studies typically focus on cost estimates and do not quantify benefits. Existing studies follow different methodologies and are difficult to compare. Many estimates of adaptation costs include investments needed to address current climate variability as well as long-term climate change. They often cost out "hard" solutions – technical and structural responses – rather than "soft" solutions, such as behavioural change, that can be less costly. Also, many of the investments needed for adaptation could take place within normal investment replacement cycles. Studies rarely split out the marginal additional costs related to climate change from those due to the broader range of pressures on water systems resulting from socio-economic drivers.

An assessment of adaptation cost and benefits by Agrawala and Fankhauser in 2008 found that overall, the literature on adapting water supply and demand in response to the impacts of climate change was still too sparse and context specific to make a broad assessment with regard to costs. However, some general insights could be drawn. For regions where precipitation is expected to increase, issues such as flood management and wastewater treatment may impose substantial additional costs. In regions with declining precipitation or shifts in the form of precipitation, investments in enhanced storage may be significant and enhancing the efficiency of water allocation becomes highly valuable. Maintaining the quality of freshwater systems may also be very costly in some cases.

A recent review of economic assessments of adaptation in Europe, "ClimateCost" (2011), found low to medium coverage of studies of the water sector. The review found that several recent national level studies imply large adaptation costs, particularly for flood protection. For example, the UK Foresight study estimated the total adaptation investment needed to address flooding (coastal, river and intra-urban) over the next 80 years at between GBP 22 billion and GBP 75 billion for a portfolio of responses, depending on the scenario, implying average annual costs of up to EUR 1 billion per year. Similarly, a recently conducted assessment on flood protection and flood risk management in the Netherlands estimates that the implementation of a comprehensive set of adaptation measures will cost EUR 1.2-1.6 billion per year up to 2050 and EUR 0.9-1.5 billion per year during the period 2050-2100. A Swedish national study estimated potentially large investment costs for adaptation across a wider range of sectors (including transport, water treatment, infrastructure, flood protection) of up to EUR 10 billion (total) in the period 2010-2100.

Although the cost and benefits of adaptation measures may be difficult to assess with precision, the costs of not adapting and/or of mal-adaptation can be particularly high, especially in a climate-sensitive domain with long-lived capital-intensive infrastructure and high sunk costs, such as water.

Source: ClimateCost (2011), "The Costs and Benefits of Adaptation Policy in Europe: Review Summary and Synthesis", *ClimateCost Policy Brief*, *www.climatecost.cc/images/Review_of_European_Costs_and_Benefits_of_Adaptation.pdf* (accessed 5 November 2012); Agrawala and Fankhauser (2008), *Economic Aspects of Adaptation to Climate Change: Costs, Benefits and Policy Instruments*, OECD Publishing, *http://dx.doi.org/10.1787/9789264046214-en*.

by the parties involved. The lack of perception of a need for action or lack of perception of a benefit from the action presents two potential obstacles to timely or effective adaptation. Timing errors are likely in either direction – either too early or too late – with implications for the economic efficiency of adaptation.

While robust approaches that perform well under a range of possible futures may be more appropriate to guide adaptation decisions than approaches seeking to optimise, **policy guidelines** (Box 2.4) can be used to prioritise action and to facilitate more timely, economically efficient, and equitable responses.

Box 2.4. **Policy guidelines to facilitate adaptation for water systems**

- **Explicitly address the risk implications of water policies.** This requires the clarification of roles and responsibilities, assigning risk to actors able to manage them most efficiently and ensuring equitable risk sharing arrangements, while taking into account environmental needs.
- **Consider the full range of strategies to manage water risks** ("avoid", "reduce", "transfer" or "bear" the risk) to address the multiple drivers of water risks including not only climate hazards, but also the socio-economic drivers affecting the exposure and vulnerability of populations, physical assets and ecosystems.
- Consider climate change adaptation **early in the planning and project cycle.** This is likely to be less costly than tacking on additional safety margins once the project has already been designed and sited and/or retro-fitting later.
- Use **appropriate decision-making approaches** to deal with pervasive uncertainty in order to improve the robustness of decisions, so that they work well under a range of possible futures.
- **Consider *expected* costs and benefits** of adaptive actions. This may require using a discount rate appropriate for long time frames.
- **Account for the option value** of approaches that allow for scalability and flexibility to favour investments that allow for adjustments as new information is gained. Irreversible and costly actions should be delayed as long as practicable.
- **Prioritise "no regrets" and "low regrets" options** (options viable under all plausible futures).
- **Minimise timing errors** by adopting a flexible approach to planning and investments under uncertainty for long-lived, water infrastructures with high sunk costs that are sensitive to changes in climate.

"Know", "target", and "manage" water risks: A role for government

In a risk-based approach, there is a need to "**know**", "**target**", and "**manage**" water risks. Governments have an important role to facilitate the provision of information to reduce information asymmetries and form the basis for making effective and informed risk management decisions ("know" the risk). Governments also have a responsibility to facilitate stakeholders' agreement on the acceptability and tolerability for a given risk in a given location ("target" the risk). Finally, governments can provide clarity in terms of the roles and responsibilities to share risk and ensure that risk is managed by actors able to do so most efficiently ("manage" the risk).

In contrast to climate change mitigation, adaptation benefits are often private and actions are largely decentralised. A distinction can be made between "private adaptation" where benefits accrue solely to the actor undertaking them, and "joint adaptation", where benefits accrue to multiple actors (Mendelsohn, 2000). Joint adaptation may be especially important for water systems, given the significant public role in provision of water supply and sanitation, flood protection and protection of freshwater ecosystems. Governments have a clear role in promoting joint adaptation, which may be under-provisioned due to its public good attributes.

Governments also have a role to play in facilitating private, or "autonomous", adaptation, by providing an enabling environment that allows private agents to make timely, well-informed, and efficient decisions or where private adaptation activities create negative externalities (Agrawala and Fankhauser, 2008). Policy responses may be necessary to overcome information failures, such as asymmetric information, which occurs when one party in an exchange has better information than another. In addition, private actors may not be fully aware of climate change impacts and the risks they entail. Government action can help to address co-ordination failures, which may arise in the case of risk-risk trade-offs. In cases where risk-risk trade-offs cross administrative boundaries, co-ordination between government actors and stakeholders is particularly important.

Governments also have a role to address institutional, regulatory, and economic barriers that may inhibit timely and efficient adaptation. Regulatory barriers may include constraints to the development of alternative water supply sources (e.g. recycled water) or restrictions on adjusting water pricing that constrain water operators' ability to recover costs related to adaptation. Lack of clarity regarding water entitlements can inhibit efficient allocation or contribute to uncertainty about future allocations. Existing building codes, standards, land-use planning and insurance schemes that are designed for a stationary climate can dull incentives (or create perverse incentives) to adapt to long-term climate change. Poorly designed economic instruments (or lack thereof) can also create barriers to adaptation. Examples include subsidies that encourage over-abstraction or inefficient use of water or subsidised flood insurance that encourage construction in flood plains.

Concluding remarks

In the absence of accurate and precise climate predictions, a risk-based approach can be used to inform adaptation decisions and to guide governments to "know", "target" and "manage" water risks. While climate change adaptation should be considered in the broader context of achieving water security and not in an isolated manner, it is also a critical aspect, even a prerequisite for achieving resilient water security over the long-term. Delaying or neglecting to systematically integrate climate change into water management policies, planning and practice can jeopardise water security and make its achievement more costly over time.

This chapter has set out a risk-based approach and policy guidelines that can be used to facilitate more timely, economically efficient and equitable adaptation for water. While there has been significant discussion and debate regarding what governments *could do* or *should do* to adapt water systems to climate change, a systematic review of what governments *are doing* is also needed to gauge progress and to gain insights from practical adaptation efforts. Chapter 3 draws on an OECD survey of policies for water and climate change adaptation across all 34 member countries and the European Commission. It summarises general trends and lessons learned, sketching a broad picture adaptation for water in practice.

Notes

1. This chapter is closely linked to "Why does water security matter?", Chapter 1 of the OECD report (2013) *Water Security for Better Lives*, which elaborates a more comprehensive discussion of the key risk terminology and concepts.
2. A "climate-driven" approach refers to a "top-down" approach that takes climate change projections as a starting point then uses downscaled projections as the basis for impact assessments used to inform adaptation policies. In contrast, "mainstreaming" is a "bottom-up" approach that integrates adaptation into plans, programs, policies and projects.
3. See, for example, Parmesan et al., (2011) for a discussion on why attempts to attribute individual impacts to rising greenhouse gases are ill advised.
4. In line with Grey and Garrick (2012), "Water security is defined as a tolerable level of water related risk at any scale and for any actor."
5. When a system is close to a tipping point, it can take a long time to recover from even a very small disturbance.
6. The temptation to select an "average" possible future for planning purposes can be strong. This is one reason why scenario developers often devise an even number of scenarios (typically four), so that there is no "average" or "middle" scenario.

References

Agrawala, S. and S. Fankhauser (eds.) (2008), *Economic Aspects of Adaptation to Climate Change: Costs, Benefits and Policy Instruments*, OECD Publishing, *http://dx.doi.org/10.1787/9789264046214-en*.

Antón, J. et al. (2012), "A Comparative Study of Risk Management in Agriculture under Climate Change", *OECD Food, Agriculture and Fisheries Papers*, No. 58, OECD Publishing, Paris, *http://dx.doi.org/10.1787/5k94d6fx5bd8-en*.

Brown, C. and R.L. Wilby (2012), "An Alternative Approach to Assessing Climate Risks", *Eos, Transactions*, Vol. 93/41, American Geophysical Union, Washington, DC., pp. 401-402. *http://dx.doi.org/10.1029/2012EO410001*.

ClimateCost (2011), "The Costs and Benefits of Adaptation Policy in Europe: Review Summary and Synthesis", *ClimateCost Policy Brief*, *www.climatecost.cc/images/Review_of_European_Costs_and_Benefits_of_Adaptation.pdf* (accessed 5 November 2012).

Coates, D. and M. Smith (2012), "Natural Infrastructure Solutions for Water Security", *in Water and the Green Economy: Capacity Development Aspects*, R. Ardakanian and D. Jaeger (eds.), *UN-Water Decade Programme on Capacity Development*, *www.unwater.unu.edu/file/get/534* (accessed 23 April 2013).

Grafton, R.Q. et al. (2012), "Global Insights into Water Resources, Climate Change and Governance", *Nature Climate Change*, Nature Publishing Group, Vol. 3, *http://dx.doi.org/10.1038/nclimate1746*.

Graham, J.D. and J.B. Wiener (1995), "Confronting Risk Tradeoffs", *in Risk vs. Risk. Tradeoffs in Protecting Health and the Environment*, pp. 1-41, Cambridge, MA, Harvard University Press.

Grey, D. and D. Garrick (2012), "Water Security as a 21st Century Challenge", *Brief*, No. 1, International Conference on Water Security, Risk and Society, University of Oxford, 16-18 April, January, *www.eci.ox.ac.uk/watersecurity/downloads/briefs/1-grey-garrick-2012.pdf* (accessed 20 October 2012).

Hall, J. (2012), "Water Security and Environmental Risk", *Brief*, No. 2, International Conference on Water Security, Risk and Society, University of Oxford, 16-18 April, March, *www.eci.ox.ac.uk/watersecurity/downloads/briefs/2-hall-2012.pdf* (accessed 20 October 2012).

Hallegate, S. et al. (2012), "Investment Decision Making under Deep Uncertainty: Application to Climate Change", *Policy Research Working Paper*, No. 6193, The Work Bank, Sustainable Development Network, Office of the Chief Economist, September, Washington, DC.

Hanemann, M. (2008), "Some Observations on the Economics of Adaptation", Presentation at the OECD Expert Workshop on Economic Aspects of Adaptation, Paris, 7-8 April, *www.oecd.org/env/cc/agendaoftheexpertworkshopontheeconomicaspectsofadaptationtoclimatechange.htm*, accessed 15 October 2012.

Huntington, T. (2006), "Evidence for Intensification of the Global Water Cycle: Review and Synthesis", *Journal of Hydrology*, Vol. 319, Elsevier, pp. 83-95, *http://dx.doi.org/10.1016/j.jhydrol.2005.07.003*.

Intergovernmental Panel for Climate Change (IPCC) (2012), "Managing the Risks of Extreme Events and Disasters to Advance Climate Change Adaptation", *A Special Report of Working Groups I and II of the*

Intergovernmental Panel on Climate Change, Cambridge University Press, Cambridge, UK and New York, NY, USA.

Klinke, A. and O. Renn (2012), "Adaptive and Integrative Governance on Risk and Uncertainty", *Journal of Risk Research*, Vol. 15/3, Routledge, London, pp. 273-292, *http://dx.doi.org/10.1080/13669877.2011.636838*.

Knight, F.H. (1921), *Risk, Uncertainty, and Profit*, Hart, Schaffner and Marx, Houghton Mifflin Company, Boston.

Loucks, D.P. (2011), "Risk and Uncertainty in Water Planning and Management: A Basic Introduction", *in* R.Q. Grafton and K. Hussey (eds.), *Water Resources Planning and Management*, Cambridge University Press, Cambridge, pp. 230-250.

Mendelsohn, R. (2000), "Efficient Adaptation to Climate Change", *Climatic Change*, 45, Kluwer Academic Publishes, the Netherlands, pp. 583-600.

Milly, P.C.D. et al. (2008), "Stationarity is Dead: Whither Water Management", *Science*,Vol. 319, American Association for the Advancement of Science, pp. 573-574, *http://dx.doi.org/10.1126/science.1151915*.

OECD (2013), "Water Security for Better Lives", *OECD Studies on Water*, OECD Publishing, *http://doi.org/10.1787/9789264202405-en*.

OECD (2009), "Innovation in Country Risk Management", *OECD Studies in Risk Management*, OECD Publishing.

Parmesan, C. et al. (2011), "Overstretching Attribution", *Nature Climate Change*, Vol. 1, Macmillan Publishers Limited.

Rees, J.A. (2002), "Risk and Integrated Water Management", *TEC Background Papers*, No. 6, Global Water Partnership, Stockholm.

Renn, O. and P. Graham (2006), *Risk Governance: Towards an Integrative Approach*, International Risk Governance Council (IRGC), *White Paper*, No. 1, *www.irgc.org/IMG/pdf/IRGC_WP_No_1_Risk_Governance__reprinted_version_.pdf* (accessed 4 December 2012).

Water and Climate Change Adaptation

Chapter 3

Climate change adaptation for water systems in OECD countries

Progress on adapting water systems to climate change has been advancing in recent years across OECD countries. In order to gauge progress and to gain insights from practical adaptation efforts for water systems, the OECD Secretariat undertook a survey of policies across all 34 member countries and the European Commission.

This chapter looks at general trends, highlights innovative approaches, and draws out lessons learned from the survey. It provides practical insights into countries' efforts to "know", "target" and "manage" increasing water risks in a changing climate. It captures the primary concerns and vulnerabilities of countries. It documents the types of policy instruments most commonly used to tackle various water challenges and identifies a range of policy levers for adaptation illustrated by practical examples. Finally, it takes stock of efforts to date to finance adaptation for water systems.

Key messages

- The OECD Secretariat carried out a **survey to take stock of recent progress** to facilitate adaptation to the effects of climate change on water systems across all 34 OECD member countries and the European Commission. It was used to capture general trends, highlight innovative approaches, and draw out lessons learned.
- **All OECD countries have already observed changes** in freshwater systems. Nearly all countries expect an increase in water risks in a changing climate. **Water shortage and extreme events (e.g. floods and droughts)** are the most frequently cited concerns.
- In the development of adaptation strategies or plans, **water** is nearly always addressed as a **priority sector or cross-cutting theme** vital for a number of key policy domains (e.g. energy, agriculture, infrastructure, biodiversity, and health). Climate change adaptation is also be being mainstreamed into existing water policies. Both approaches are important to ensure coherence and effectiveness.
- The majority of efforts to date have focussed on "knowing" the risk by **building the scientific evidence base** and disseminating information. Given the significant investments being made to improve the scientific evidence base, there is an urgent need to ensure that this evidence is used to best effect and meets the needs of users making practical, on-site adaptation decisions.
- In terms of **policy responses, information-based instruments** (e.g. flood risk maps, decisions support tools for risk management, adaptation guidance for local governments) are by far the most widely used.
- Some countries are also revising **laws and regulations** (e.g. sustainable water abstraction limits, building codes, land-use planning) and adjusting economic instruments (e.g. water tariffs, water-related environmental taxes, flood insurance schemes) to reduce baseline stress on water systems, raise financing and address increasing flood risks.
- Only a handful of countries have begun to address the issue of **financing adaptation**. Some countries are mainstreaming adaptation into existing budgetary mechanisms, while others are addressing adaptation via specific water programmes or projects. Some countries are tapping international financing mechanisms. A few countries have allocated dedicated funding to climate change adaptation in general, which typically includes measures for water.

Progress on adapting water systems to climate change has been advancing in recent years across OECD countries. To develop an up-to-date information base on how countries are addressing this issue in practice, the OECD undertook a survey to take stock of recent experience. A country profile has been prepared for each member country and the European Commission.[1] Each country profile provides (where the relevant information is available):

- An **overview of the main observed trends and projected impacts** of climate change on freshwater, an indication of primary concerns, and key vulnerabilities.
- A list of the **key policy documents** relevant to water and climate change adaptation.
- An inventory of the main **regulatory, economic, and information-based instruments** (implemented or under development) used to facilitate adaptation for water systems.
- A summary of the **main research programmes** focussed on climate change and water.
- A summary of the **principal financing mechanisms** and **investment programmes**.
- Examples of **innovative initiatives**.

The country profiles cover measures to address impacts on the quantity and quality of water resources, water supply and sanitation, extreme events (e.g. floods and droughts), and freshwater ecosystems.

The information captured in the country summary covers the main elements of climate change adaptation activities for water (Figure 3.1). The survey of OECD countries demonstrates that countries take different approaches to deal with adaptation that do not necessarily include all of these generic elements, nor follow a specific sequence. Typically, countries start with building the scientific evidence base, which informs the development of strategies and plans. Measures identified in plans or programmes are then implemented. While many countries have developed adaptation responses at the national level, some have taken a sub-national and/or sectoral approach.

Figure 3.1. **Generic building blocks for climate change adaptation for water**

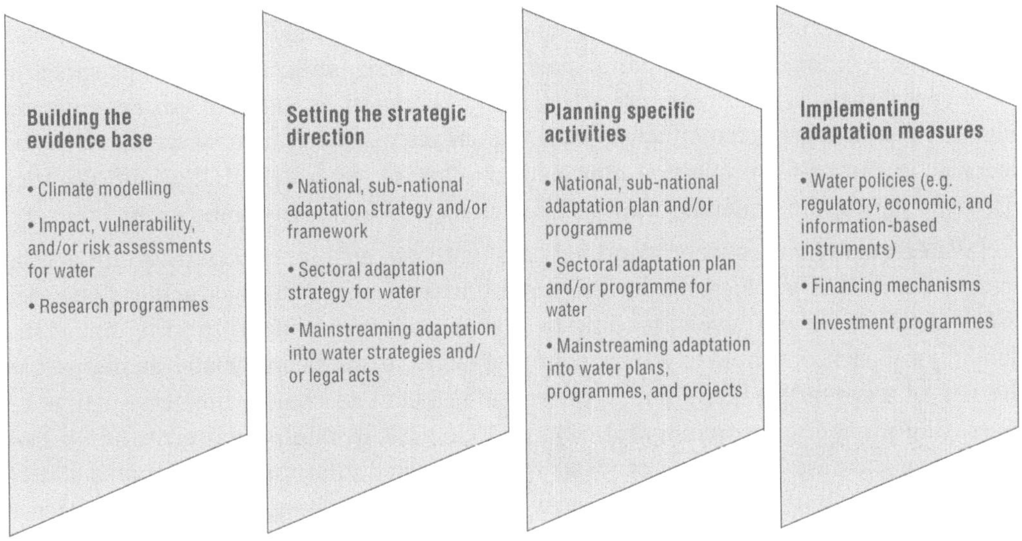

The country profiles are not intended to be a comprehensive inventory of all possible measures that may facilitate climate change adaptation for water in each country. They are meant to provide an accurate and up-to-date illustration of the *main* projected impacts, policy responses and innovative approaches and to direct readers to the relevant references. To serve as a tool for information sharing, this descriptive information base has been compiled in an annex to this report (online only). It has also been used to draw out general trends, lessons learned, and innovative approaches, which are summarised in this chapter.

Building the evidence base to "know" the risk

In recent years, OECD countries have improved the evidence base for understanding climate change impacts on water systems and invested significantly in research capacity. Most countries have identified one or more areas of concern relating to water quantity, water quality, water supply and sanitation, extreme events (e.g. floods and droughts) and freshwater ecosystems (Table 3.1). Around half of countries have already undertaken comprehensive national level impact, vulnerability, or risk assessments, in which impacts on water are often a key focus. Many countries have established research programmes or projects to further develop the scientific evidence base on impacts on water systems and possible adaptation responses. This section provides an overview of OECD countries' efforts to "know" the risk.

Trends, projections, primary concerns and vulnerabilities

Nearly all OECD countries have observed shifts in trends relating to water and project increasing water risks in the future. These changes include precipitation changes, shifts in river flow and discharge, changes in glaciers, snow and ice, increases in air and water temperature, water quality degradation, increased frequency and magnitude of extreme events, such as floods and droughts, and impacts on freshwater ecosystems. The most frequently cited concerns are increasing water shortage and increasing frequency and intensity of extreme events, especially floods and droughts.

Water quantity was flagged as a priority issue in 23 (of 34) OECD countries and by the European Commission. Countries with arid climates, such as Greece, Israel, Spain, Turkey, along with Southwest Australia and the Southwest US, can be especially sensitive to even small changes in precipitation. Even countries considered relatively water abundant overall, such as France or the Netherlands, anticipate increased water stress in vulnerable regions. In some cases, seasonal shortages are the source of concern. In Belgium, for example, an increase in winter precipitation is expected to increase water recharge, while the expected decrease in summer precipitation may lead to a lack of water availability, with negative effects for agriculture, waterway transport, aquatic recreation, tourism and so on.

Water quality issues were identified as a main concern by 15 countries. In Canada, warmer conditions will increase surface water temperatures, decrease the duration of ice cover and lower water levels, leading to higher pollutant concentrations. Increased flooding will also contribute to water quality degradation. Korea anticipates an increase in the risk of algae outbreaks in public waters due to climate change impacts such as an increasing number of scorching hot days and changes in rainfall patterns. Korea also expects an increase in the risk of water quality degradation due to non-point source pollution resulting from increasing quantity and intensity of precipitation. Chile expects a range of water quality issues related to climate change. In Chile, the degradation of surface water quality due to extreme events is projected. Groundwater salinisation in coastal zones is expected due to the intrusion of salt water, while degradation of groundwater quality is expected in northern areas. The capacity of water systems to dilute pollutants is also expected to decrease.

Table 3.1. **Primary concerns in OECD countries and the EC**

<table>
<tr><th rowspan="2">Country / Area of concern</th><th rowspan="2">QN</th><th rowspan="2">QL</th><th rowspan="2">WSS</th><th colspan="2">Extreme weather events</th><th rowspan="2">ECO</th></tr>
<tr><th>Flood</th><th>Drought</th></tr>
<tr><td>Australia</td><td>●</td><td>●</td><td>●</td><td>●</td><td>●</td><td>●</td></tr>
<tr><td>Austria</td><td>●</td><td></td><td>●</td><td></td><td>●</td><td>●</td></tr>
<tr><td>Belgium</td><td>●</td><td>●</td><td>●</td><td>●</td><td>●</td><td>●</td></tr>
<tr><td>Canada</td><td>●</td><td>●</td><td></td><td colspan="2">●</td><td></td></tr>
<tr><td>Chile</td><td>●</td><td>●</td><td>●</td><td>●</td><td>●</td><td>●</td></tr>
<tr><td>Czech Republic</td><td>●</td><td>●</td><td></td><td>●</td><td>●</td><td></td></tr>
<tr><td>Denmark</td><td></td><td></td><td>●</td><td colspan="2">●</td><td></td></tr>
<tr><td>Estonia</td><td>●</td><td>●</td><td>●</td><td>●</td><td></td><td>●</td></tr>
<tr><td>Finland</td><td></td><td></td><td></td><td colspan="2">●</td><td></td></tr>
<tr><td>France</td><td>●</td><td></td><td></td><td></td><td>●</td><td></td></tr>
<tr><td>Germany</td><td>●</td><td></td><td></td><td>●</td><td>●</td><td></td></tr>
<tr><td>Greece</td><td>●</td><td></td><td></td><td></td><td>●</td><td></td></tr>
<tr><td>Hungary</td><td></td><td></td><td></td><td>●</td><td>●</td><td></td></tr>
<tr><td>Iceland</td><td></td><td></td><td></td><td></td><td></td><td></td></tr>
<tr><td>Ireland</td><td></td><td></td><td>●</td><td></td><td></td><td></td></tr>
<tr><td>Israel</td><td>●</td><td>●</td><td></td><td colspan="2">●</td><td></td></tr>
<tr><td>Italy</td><td>●</td><td></td><td></td><td></td><td>●</td><td>●</td></tr>
<tr><td>Japan</td><td></td><td></td><td></td><td colspan="2">●</td><td></td></tr>
<tr><td>Korea</td><td>●</td><td>●</td><td>●</td><td>●</td><td></td><td>●</td></tr>
<tr><td>Luxembourg</td><td>●</td><td>●</td><td>●</td><td>●</td><td></td><td></td></tr>
<tr><td>Mexico</td><td>●</td><td></td><td>●</td><td>●</td><td>●</td><td>●</td></tr>
<tr><td>Netherlands</td><td>●</td><td></td><td></td><td>●</td><td>●</td><td></td></tr>
<tr><td>New Zealand</td><td></td><td></td><td></td><td>●</td><td>●</td><td></td></tr>
<tr><td>Norway</td><td></td><td></td><td></td><td colspan="2">●</td><td>●</td></tr>
<tr><td>Poland</td><td>●</td><td>●</td><td>●</td><td>●</td><td>●</td><td>●</td></tr>
<tr><td>Portugal</td><td>●</td><td></td><td></td><td>●</td><td>●</td><td></td></tr>
<tr><td>Slovak Republic</td><td></td><td></td><td></td><td>●</td><td>●</td><td></td></tr>
<tr><td>Slovenia</td><td>●</td><td>●</td><td>●</td><td>●</td><td>●</td><td></td></tr>
<tr><td>Spain</td><td>●</td><td>●</td><td>●</td><td>●</td><td>●</td><td>●</td></tr>
<tr><td>Sweden</td><td></td><td></td><td></td><td>●</td><td></td><td></td></tr>
<tr><td>Switzerland</td><td></td><td>●</td><td></td><td colspan="2">●</td><td>●</td></tr>
<tr><td>Turkey</td><td>●</td><td>●</td><td>●</td><td>●</td><td>●</td><td>●</td></tr>
<tr><td>United Kingdom</td><td>●</td><td></td><td>●</td><td>●</td><td></td><td></td></tr>
<tr><td>United States</td><td>●</td><td>●</td><td>●</td><td></td><td>●</td><td></td></tr>
<tr><td>European Commission</td><td>●</td><td></td><td></td><td colspan="2">●</td><td></td></tr>
</table>

Note: QN = quantity; QL = quality; WSS = water supply and sanitation; ECO = freshwater ecosystems.
In cases where the box under the extreme weather events is merged, countries have indicated extreme weather events as primary concern, but not specified whether this relates to flood, drought, or both.
Source: See country profiles associated with this publication on iLibrary as well as at *www.oecd.org/env/resources/waterandclimatechange.htm.*

Nearly all OECD countries (32) as well as the European Commission consider **extreme events** a primary concern. Both **floods and droughts** are a primary concern for a number of countries, including Australia, Belgium, Chile, the Czech Republic, Germany, Hungary, Mexico, the Netherlands, New Zealand, Poland, Portugal, the Slovak Republic, Slovenia, Spain and Turkey. For example, the Netherlands is highly vulnerable to floods with around 24% of the land below sea level. Sixty per cent of Dutch territory is vulnerable to flooding from either the sea or the rivers without water defences. At the same time, further warming and an increasing precipitation deficit could cause considerable problems for freshwater supply in the Netherlands as early as 2050. The adaptive capacity of the freshwater supply is considered limited in its current setting. Both floods and droughts are also a primary concern in Germany. Eastern Germany is vulnerable to seasonal drought risk, with potential impacts on agriculture and forestry while some river catchments (e.g. Rhine, Danube, Elbe, and Oder) are vulnerable to flood risk.

Threats to **freshwater ecosystems** were flagged as a significant concern for 13 OECD countries, including Australia, Austria, Belgium, Chile, Estonia, Italy, Korea, Mexico, Norway, Poland, Spain, Switzerland, and Turkey. In Australia, increases in drought, salinity and other water quality issues, as well as changes in water availability, are threats to ecosystems. In Norway, climate change is expected to change the distribution of species by moving species towards areas of higher altitude and latitude. Some species of fish and crustaceans may be unable to migrate to alternative habitats due to isolation between freshwater systems. Water temperature can rise above critical levels for important fish species like salmon, trout and charr in some areas, such as rivers in southern Norway. Regulated rivers with low minimum water flow are considered at the highest risk. In the US, a warmer climate in the southwest is predicted to result in widespread tree mortality that will likely cause substantial changes in forest and species distributions, altering watershed hydrology.

In some cases, impacts on the water cycle may be projected, but do not necessarily translate into areas of concern requiring adaptation. For example, Germany anticipates more frequent occurrence of low-water periods, but it does not expect fundamental drinking water supply problems, since drinking water supplies mostly make use of local groundwater resources and only draw on surface water to a limited extent. The European Commission has indicated that in Eastern Europe, no major water user is particularly threatened and water scarcity can be reduced through integrated water management.

Some countries have identified **positive effects of climate change on water systems**. For example, groundwater recharge is projected to increase in Estonia. The magnitude of the increase will depend on the specific hydro-geological conditions of catchments. Increased groundwater recharge is expected to be the most intensive in the Pandivere Upland, which is the most important groundwater catchment area in the country. In Austria, shifts in the seasonal precipitation distribution and enhanced water runoff in winter due to a higher fraction of rainfall (rather than snow) are expected to lead to a more uniformly distributed annual cycle of water run-off. In general, this favours the production of hydropower, as supply will be more in phase with electricity demand. However, possible increases in sediment loading would impair the operation of hydropower infrastructure.

Yet, even where projected impacts are expected to be largely positive, climate change impacts on water systems may still pose challenges that require adaptation. For instance, in Iceland where freshwater supplies are abundant, changes in the courses of glacial rivers due to the rapid retreat of glaciers may affect roads and other lines of communication. The rapid retreat of glaciers is also projected to cause major changes in runoff, which may have practical implications for the design and operation of hydroelectric power plants.

Areas of acute vulnerability are seeing earlier and more significant climate change impacts that require urgent attention. For example, Canada's Arctic is particularly vulnerable to climate change. Infrastructure in the Arctic regions has already been exposed to significant impacts that require urgent changes to engineering practices, climate-related design parameters, codes and standards from melting permafrost, warming temperatures, changing snow loads and other extreme precipitation.

Risk appraisal: Impact, vulnerability and risk assessments

Among 34 OECD countries, around half have already undertaken comprehensive national level impact, vulnerability, or risk assessments, in which impacts on water are nearly always addressed. In the UK, a climate change risk assessment (CCRA) is a statutory requirement of the Climate Change Act of 2008 (Box 3.1). The first CCRA was released in January 2012. In the US, the most recent iteration of the National Climate Assessment is currently underway. It is expected to be complete in 2013.[2] The assessment is being conducted under the auspices of the Global Change Research Act of 1990, which requires a report to the President and Congress every four years. Assessments are currently under development in many other countries. For example, Denmark, Estonia, Israel, and Poland are currently developing national level impact, vulnerability, or risk assessments.

Box 3.1. **The UK Climate Change Risk Assessment (CCRA)**

The UK is the first country in the world to build a risk-based approach to climate change into legislation. The Climate Change Act of 2008 specifies that a Climate Change Risk Assessment (CCRA) must be carried out every five years. The first CCRA was published by the Government on 25 January 2012.

Acknowledging that the future is highly uncertain, the assessment uses the best scientific evidence available alongside well-established risk-based decision-making approaches to assess risks and to decide how to respond. The CCRA reviewed the evidence for over 700 potential impacts of climate change in the UK context. Detailed analysis was undertaken for over 100 of these impacts across 11 key sectors, on the basis of their likelihood, the scale of their potential consequences and the urgency with which action may be needed to address them. Findings from the CCRA were used to set out the main priorities for adaptation in the UK.

Several of the most significant findings of the CCRA relate to water. The risk of flooding is projected to increase significantly across the UK. The expected annual damage to properties caused by flooding from rivers and the sea is currently approximately GBP 1.3 billion per annum for the UK as a whole and GBP 1.2 billion for England and Wales.[1] New analysis for England and Wales showed that future potential risk estimates are within the following ranges: GBP 1.5 billion to GBP 3.5 billion by the 2020s, GBP 1.6 billion to GBP 6.8 billion by the 2050s, and GBP 2.1 billion to GBP 12 billion by the 2080s.[2]

Another significant CCRA finding is that the UK's water resources will be under increasing pressure due to changes in hydrological conditions, population growth and regulatory requirements to maintain good ecological status. Major supply-demand deficits for public water supplies are projected in five river basin regions – Anglian, Humber, Severn, North West England and the Thames basins. The Thames river basin, which provides the current water supply to London, is estimated to face the largest deficits based on a central population projection. Planned improvements in water efficiency and new supply schemes are likely to be sufficient to manage risks in the near term (2020s) but the widening supply-demand gap presents a considerable challenge for the medium term (2050s). The potential risks are greatest in England and Wales and may affect people through changes in the service offered by water companies, changes to the costs of water and the environmental quality of rivers and lakes.

Box 3.1. **The UK Climate Change Risk Assessment (CCRA)** *(cont.)*

The CCRA also found that sensitive ecosystems that are already threatened by land use changes may be placed under increasing pressure due to climate change. The main direct impacts relate to changes in the timing of life cycle events, species distribution and ranges and potential changes in hydrological conditions that may affect aquatic habitats.

1. Note that accurate estimates are only available for England and Wales so the UK wide estimate is approximate and considers the additional numbers of properties at risk and population estimates for Scotland and Northern Ireland.
2. These estimates assume continued investment to maintain the condition of existing flood defences but do not include other flood risk management measures. Future risks of flooding will depend upon the location and pattern of future development and level of additional investment in flood risk management (by government and local communities), as well as changes to the hydrological cycle and rates of sea level rise. Future targeted investment may substantially reduce these damage costs. These estimates include increases in population.

Source: Department for Environment, Food and Rural Affairs (DEFRA) (2012), *UK Climate Change Risk Assessment: Government Report*, DEFRA, London, *www.gov.uk/government/publications/uk-climate-change-risk-assessment-government-report* (accessed 19 November 2012).

Countries have also conducted assessments at the sub-national level or have taken a sectoral approach. In Belgium, impact, vulnerability and adaptation assessments have been funded and piloted at regional level (in Wallonia and in Flanders). Most of Spain's Autonomous Communities have developed climate change adaptation strategies or plans. In addition, Spain undertook a national assessment of the impact of climate change on water resources at the national level in 2010. Austria has carried out vulnerability assessments for nine sectors including water, agriculture, and ecosystems and biodiversity, and so on. Slovenia has developed preliminary risk assessment for floods.

Other assessments have taken a specific regional or geographic focus. For instance, given the vulnerability of alpine regions to climate change, the vulnerability assessment undertaken by the European project "Climate Change Adaptation by Spatial Planning in the Alpine Space" (CLISP) focuses on how spatial planning can reduce vulnerability to climate change in alpine regions. The project is led by the Austrian Federal Environment Agency and involves 14 partners from four other countries, including Italy, Germany, Slovenia and Switzerland. Luxembourg is participating in the regional Flood and Low Water Management Moselle-Sarre project to assess the consequences of climate change for floods and low water flow in the Moselle and Saar catchments and to develop adaptation strategies.

Improving the evidence base and research capacity

Improving the scientific evidence base through research is an area where most OECD countries and the European Commission have made significant effort. Across the OECD, there is a range of ongoing research programmes addressing various water issues related to climate change. Some countries have allocated significant public funding to these efforts.

While many countries depend on international mechanisms such as IPCC climate models to develop the national climate scenarios, some OECD countries have developed their own downscaled climate models to provide higher spatial resolution of projected impacts on water systems. In France, future climate change is simulated using the French regional climate change models developed by the National Centre for Meteorological Research and the Institute Pierre-Simon Laplace. Efforts to improve climate science are also on-going at the regional level. The European Commission has invested significant funds in climate system studies and modelling under both Framework Programme 6 and 7 (Box 3.2).

Box 3.2. **Assessing vulnerability and adaptive capacity in Europe's river basins**

An integrated assessment of the vulnerability of Europe's river basins to climate change impacts and the development of adaptation measures for the water sector is currently underway. The project, Climate Adaptation – Modelling Water Scenarios and Sectoral Impacts (ClimWatAdapt), aims to shed light on both vulnerability and adaptive capacity in different sectors and across Europe's river basins. The project looks to enhance the quality of adaptation measures, the knowledge base, and facilitate the exchange of adaptation best practice between countries and regions by putting in place an integrated assessment framework. This integrated assessment framework will allow for the identification of which regions in Europe are potentially vulnerable to climate change and which adaptation measures could potentially be promoted at the EU level. The integrated framework will consist of: *i)* scenarios on climate change and socio-economic developments; *ii)* vulnerability indicators; *iii)* an inventory of measures; *iv)* instruments and methods to assess the performance of adaptation options; and *v)* a decision support tool. It covers the 27 EU member States.

Source: ClimWatAdapt (2012), "Climate Adaptation: Modeling Water Scenarios and Sectoral Impacts", *www.climwatadapt.eu* (accessed 5 November 2012).

In the UK, the most recent climate change projections are probabilistic, allowing for a measure of uncertainty to be included. These projections, known as UKCP09, are the fifth generation of climate change information in the UK. They reflect significant advances in climate science and computer modelling. The projections show a range of possible future climates and their associated probabilities. The probabilities are created by weighting future climate projections based on how well they represent the past climate. In this way, they can be seen as the relative degree to which each climate outcome is supported by the evidence available.

In developing the evidence base for climate impacts on water systems, some countries have conducted research on water in the context of broader work on climate change, others have mainstreamed the climate change adaptation in general water research. There are also many examples of research programmes dedicated specifically to climate change adaptation and water. For instance, the Water Resources and Freshwater Biodiversity Network under Australia's National Climate Change Adaptation Research Facility (NCCARF) brings together Australia's top water scientists to understand the risks to Australia's surface and groundwater resources and freshwater biodiversity due to climate change.

Research programmes are also focussing on tackling key vulnerabilities related to water. For example, glacial retreat due to climate change will have a significant impact on water supply in Chile. Glaciers act as strategic water reserves in Chile, not only by supplying water to river basins in summer, but by providing the single most important source of replenishment of rivers, lakes and groundwater in arid regions during periods of drought. To tackle this issue, the Glaciology and Snow Unit was created in 2008 within the Ministry of Public Works' General Directorate of Water. The Unit will establish and implement a national glaciology program that will develop a glacier inventory and study and monitor glaciers in Chile.

Many countries have dedicated significant funding to improve their understanding of climate change adaptation. For example, the Government of Canada invested CAN 85.9 million over the period from 2007 to 2011 to increase capacity to adapt to climate change. An additional CAN 148.8 million was allocated for the 2012-17 period. Australia has invested AUD 44 million to establish the Climate Adaptation National Research

Box 3.3. Exploring the water balance in France in 2070 and testing adaptation strategies

In order to provide a systematic view of climate change impacts related to the water cycle in France by 2070 and to inform the development of adaptation strategies, the Ministry for Ecology, Sustainable Development and Energy (MEDDE) has recently concluded a two-year study engaging hundreds of experts. The study, *Explore 2070*, aimed to anticipate the major challenges related to climate change impacts on the water cycle, prioritise risks and select adaptation responses to meet these challenges and minimise these risks. The results provide decision support for government officials and local representatives of the central government, serve as a reference for the scientific community, and provide a source of information and a tool to raise the awareness of the general public. A range of water issues were addressed including: the supply and demand of water resources, the preservation of aquatic ecosystems, protection of coastal areas, risks generated by climate hazards, integrated water management, adaptation of economic sectors linked to water.

Explore 2070 has taken an original approach by engaging a diverse group of experts from MEDDE, the National Office of Water and Aquatic Environments (ONEMA) and Water Agencies to develop climate scenarios and evaluate various adaptation hypotheses. The project developed a baseline scenario to provide a reference to compare the expected impacts of climate change in 2070 against the present day situation (represented by 2006) and to assess three possible adaptation strategies. The adaptation strategies were used to evaluate the extent to which different combinations of adaptation measures could offset negative climate impacts on the water balance. The results compared the percentage of unmet demand in 2006, projections of the percentage of unmet demand for the baseline in 2070, and each of the three proposed adaptation strategies in 2070. Projections were modelled for a "normal" year and a "dry" year (occurring once every five years). The projections for unmet demand were examined for each major group of water users – households, ecosystems, industry, energy and agriculture. The location and severity of projected water deficits were mapped.

To aid the interpretation of the results, the principle uncertainties were characterised. Uncertainties arise from incomplete knowledge and lack of data on a range of relevant parameters, including: the interaction of surface water and aquifers, land use and the effect of vegetation, operation of water infrastructures (e.g. dams, reservoirs), inter-basin transfers, and the quality of surface and groundwater. The potential impact of these uncertainties on the modelling results and the level of confidence in the results were also estimated.

Source: Ministère de l'Écologie, du Développement durable et de l'Énergie (MEDDE) (2011), *Eau et changement climatique : quelles stratégies d'adaptation possibles*, MEDDE, Paris.

Flagship under the Commonwealth Scientific and Industrial Research Organization (CSIRO) and the National Climate Change Adaptation Research Facility (NCCARF) has committed AUD 10 million over 5 years to its eight Adaptation Research Networks. A transnational project in the North Sea Region, CLIWAT, focuses on climate change impacts on groundwater and through this on surface water and water supply. It has a total budget of EUR 5.5 million, half of which is financed by the European Regional Development Fund (ERDF). Its results will inform the design of adaptive measures and new standards.

Policy frameworks to set the strategic direction

Significant progress has been made in recent years in OECD countries and by the European Commission to develop a policy response to facilitate climate change adaptation for water systems. A National Adaptation Strategy (NAS) has already been developed by 19 OECD countries (just over half) and an additional 8 countries are currently in the process of doing so. Of the countries that already have an NAS, 9 have gone on to develop a National Adaptation Plan (NAP), in addition to Luxembourg. NAPs are currently being developed by 10 countries (see the timeline in Figure 3.2).

Figure 3.2. **Timeline of development of National Adaptation Strategies and Plans in OECD countries**

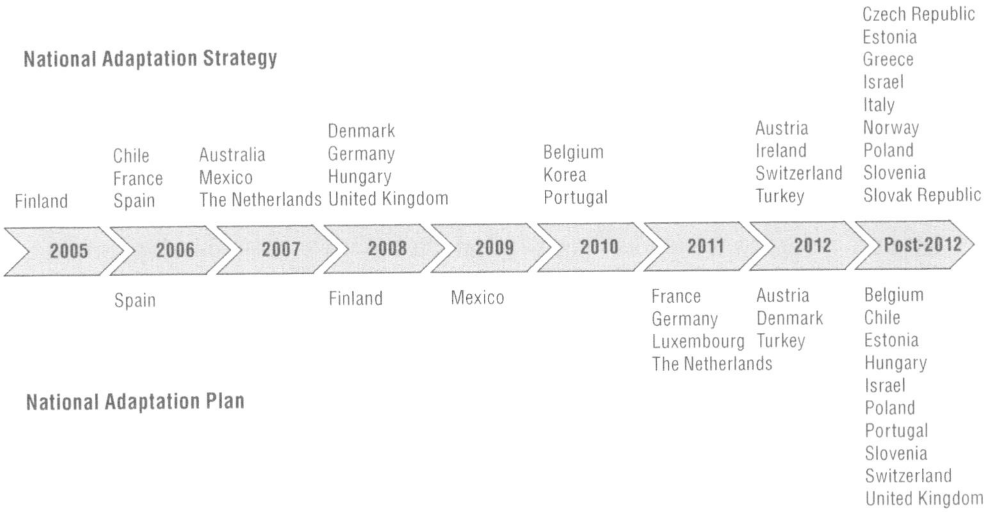

Source: See country profiles associated with this publication on iLibrary as well as at *www.oecd.org/env/resources/waterandclimatechange.htm.*

Countries almost always address water as a priority sector or thematic area within these strategies and plans. At the same time, climate change adaptation is also being actively mainstreamed into existing water policies. A significant number of countries have identified water issues, such as consecutive droughts and increased frequency and intensity of flooding events as a key driver for policy action on climate change adaptation. For example, in the wake of five consecutive years of drought that significantly reduced freshwater reserves, Israel decided to include climate change within the framework of its strategic program for the water sector.

Water as a priority in adaptation policies

Since 2005, when Finland released its National Adaptation Strategy, progress on developing strategies and plans for adaptation in OECD countries has advanced quickly. Significant progress has been made to date and a number of efforts are currently underway. For instance, Poland is currently preparing a National Strategy for Adaptation to Climate Change, which is planned for release at the end of 2013. Estonia is preparing a National Adaptation Strategy, targeted for release in 2016. Chile is currently developing an adaptation strategy for water resources, scheduled to be ready by 2015.

Turkey, for example, has undertaken substantial adaptation planning at the national level, developing a strategy and detailed action plans, which feature water resources management, natural disaster risk management and ecosystem services as priority areas. Turkey released a National Climate Change Adaptation Strategy and Action Plan in 2012. Adaptation actions include mainstreaming measures to tackle the impact of climate change on water resources in the Development Plans and Programmes, developing hydrological drought studies to better inform water resources management, and developing integrated river basin management plans that explicitly take into account impacts of climate change.

Consistent with the high priority that it has given to climate change adaptation, Spain was the first country to develop both a National Adaptation Strategy and Plan. The National Climate Change Adaptation Plan (PNACC) was adopted in 2006 and serves as the overarching framework guiding the government's response to address climate change adaptation. The First and Second PNACC Work Programmes, adopted in 2006 and 2009, respectively, developed a national programme for the assessment of impacts and vulnerabilities and specific lines of action for adaptation. Water resources are considered both a key sector as well as a cross-cutting theme (CLIMATE-ADAPT, 2012).

France has also articulated a comprehensive National Adaptation Plan addressing water resources and natural disasters as priority areas. The French National Adaptation Plan, released in 2011, was mandated by the Grenelle Law in 2009. To develop the plan, the Ministry of Ecology set up a national consultation group with diverse stakeholders. The national approach combined with local consultations resulted in 211 recommendations, and from these, 230 concrete measures were designed. To cope with uncertainty the plan gives priority to "low regret" and knowledge improvement measures. It covers a five-year period (from 2011 to 2015), after which it will be reviewed and revised. The plan takes a mainstreaming approach in order to facilitate implementation. Key adaptation measures for water include improving understanding of the impacts of climate change on water resources and the potential of adaptation scenarios, developing water saving and ensuring more efficient water use, and reinforcing the integration of climate change into water planning and management.

Many federal countries, including Canada, Belgium and New Zealand, take a decentralized approach to adaptation policy responses. In Canada, the *Federal Adaptation Policy Framework* defines the federal role as facilitating adaptation undertaken by sub-national governments and the private sector, in addition to integrating adaptation into federal policy and planning. The federal government facilitates adaptation by generating and sharing knowledge, building adaptive capacity to respond and helping individuals and businesses to adapt by providing decision-making tools. Canadian provincial authorities have been active in developing adaptation strategies and plans. For example, the Government of Ontario has developed *Climate Ready: Ontario's Adaptation Strategy and Action Plan* in 2011. All of Canada's 10 provinces and 3 territories have addressed climate change adaptation for water in their policy frameworks. Most sub-national governments have either a climate change adaptation strategy and/or a plan, and others take climate change adaptation into account in their water strategies and/or plans.

Mainstreaming adaptation into water policies

It is well recognised that, in general, sound water management will go a long way towards enhancing resilience to climate change impacts. Sound water policy can reduce baseline stress on water systems and improve adaptation to current climate variability, thus enhancing resilience to the future changes. In practice, what is considered (or labeled) as "adaptation" to climate change varies among countries. Existing water policies are also being adjusted to explicitly account for climate change. Taking a pragmatic approach, the review of country activities has included those water policies and recent reforms that countries' policy documents refer to as facilitating climate change adaptation, even if, in many cases, climate change adaptation may not be the main motivation for these initiatives.

Examples of mainstreaming climate change adaptation into water policy include the EU Guidance document on adaptation to climate change in water management produced in 2009 to ensure that the River Basin Management Plans (RBMP) are climate-proofed. Greece's Water Scarcity and Drought Management Strategies are addressing the uneven spatial and seasonal distribution of water resources more effectively than in the past in order to adapt to prolonged droughts. France's 2006 Law on Water and Aquatic Ecosystems (LEMA) expressly integrated climate change adaptation in the management of water resources. The Australian Government has recently undertaken a comprehensive assessment of the interactions between water policies and climate change adaptation policies (Box 3.4).

Box 3.4. **Assessing the links between water policies and climate change adaptation policies in Australia**

Australia has identified water management and climate change as two of the most important public policy issues facing the country, recognising that climate change poses an additional challenge to water management. The National Water Commission's 2011 biennial assessment of progress under the National Water Initiative notes that policies and investment decisions involving climate change and water are intrinsically intertwined and decisions in any of the areas can have strong impacts on the other.

A comprehensive assessment of the interactions between climate change policy and water policy was undertaken across seven key sectors that supply water, use water or otherwise affect water policy (urban water, rural water, the environment, agriculture, electricity generation, forestry and mining). The results were presented in the recent paper *Water Policy and Climate Change in Australia* (2012). One of the key findings of the assessment is that many existing water policy tools constitute, or influence, adaptation responses. While some water policies may be less synergistic with adaptation policies, many aspects of current water policies – water entitlements, market-based mechanisms for the release of unallocated water, and the use of water markets to reallocate water to different uses – provide a solid base for managing the potential climate change impacts on water systems.

Sources: The National Water Commission (2012), *Water Policy and Climate Change in Australia*, NWC, Canberra, *http://archive.nwc.gov.au/library/topic/markets/water-policy-and-climate-change-in-australia* (accessed 18 November 2012); The National Water Commission (NWC) (2011), *The National Water Initiative: Securing Australia's Water Future: 2011 Biennial Assessment*, NWC, Canberra, *http://nwc.gov.au/publications/topic/assessments/ba-2011* (accessed 18 November 2012).

In the US, the Environmental Protection Agency (EPA) is actively reviewing the implications of climate change in the context of its water programmes. While the EPA recognises that climate change has increased the importance of many of the programs and activities already underway in the National Water Programme, it has concluded that climate change poses such significant challenges to the nation's water resources that transformative approaches will be necessary. Therefore, the EPA is looking to revise data collection, analytical methods, and regulatory practices that have been developed over the past 40 years since passage of the Clean Water Act and the Safe Drinking Water Act.

Policy instruments to "know", "target" and "manage" water risks

Regulatory, economic and information-based instruments all have a role to play in effective, timely and efficient climate change adaptation for water. Table 3.2 provides examples of these three types of water policy instruments. This section provides an overview of the measures (implemented or under consideration) that countries have identified as facilitating adaptation for water systems in the OECD survey. In the section below, the counts signalling the number of countries using regulatory approaches and economic instruments do not include Canada, which is taking a decentralised approach.[3]

Table 3.2. **Examples of selected water policy instruments**

Regulatory instruments (command and control)	Economic instruments	Information and other instruments
Norms and standards for water quality (e.g. drinking water quality, ambient water quality for recreational water bodies, industrial discharges)	Charges (e.g. abstraction, pollution) user tariffs (e.g. for water services) payment for watershed services	Metering of water usage eco-labeling and certification (e.g. for agriculture, water-saving household appliances)
Performance-based standards	Reform of environmentally-harmful subsidies (e.g. production-linked agricultural support; energy subsidies for pumping water)	Voluntary agreements between businesses and government for water efficiency
Restrictions or bans on activities that have an impact on water resources (e.g. polluting activities in catchment areas, ban on phosphorus detergents)	Subsidies (e.g. public investment in infrastructure, social pricing of water)	Awareness raising and training in ecological farming practices or improved irrigation technologies
Abstraction and discharge permits Water rights	Tradable water rights and quotas	Stakeholder initiatives and co-operative arrangements seeking to improve water systems (e.g. between farmers and water utilities)
Land-use regulation and zoning (e.g. buffer zone requirements for pesticides application)	Insurance schemes	Planning tools (e.g. integrated river basin management plans) Cost-benefit analysis of water management policies

Source: OECD (2012), *OECD Environmental Outlook to 2050: The Consequences of Inaction*, OECD Publishing, *http://dx.doi.org/10.1787/9789264122246-en*.

Regulatory instruments

The OECD survey indicates that at least 26 countries, along with the European Commission, have identified regulatory instruments as part of their adaptation response for water systems. Regulatory approaches are mainly being used to cope with water quantity issues and extreme events, but also being used to address issues related to water quality and water supply and sanitation. Measures include reviewing existing legislation, regulations and plans and updating standards in order to explicitly address climate change impacts. Finland, Germany, and Switzerland, for example, are among the countries conducting reviews of existing water legislation to consider climate change impacts. Slovakia is re-evaluating usable groundwater reserves, taking into account climate change impacts. Turkey is stepping up efforts to prevent the illicit use of groundwater resources. Table 3.3 provides additional specific examples.

In some cases, new measures have been developed in response to specific climate change concerns. In May 2012, the Danish parliament passed a law, which provides municipalities with a legislative basis to directly account for climate change adaptation in local city planning decisions. The new law allows municipalities to ban construction in certain areas solely due reasons relating to climate change adaptation.

Table 3.3. **Examples of regulatory measures to address water quantity issues**

Australia	• The Murray-Darling Basin Plan sets enforceable limits on the use of water that reflect an environmentally sustainable level of use.
Belgium	• "*Water toets*" requirements (Flanders) scrutinises building plans for their potential impacts on water, particularly in flood-prone areas or near drinking water catchment areas.
Finland	• Review of existing permits for each catchment to determine how effective they are in responding to climate change impacts on water conditions.
France	• New procedure for determining abstractable volumes to ensure that the demand does not exceed the local supply in sub-basin areas.
Slovak Republic	• Re-evaluation of usable groundwater reserves to achieve good status, taking into account climate change impacts.
Spain	• Statutory requirement for water plans and the inventory of water resources to assess climate change impacts.
Switzerland	• Review of the existing legislation with regard to minimum low flows, lake regulation and management.
Turkey	• Preventing illicit use of groundwater resources.
United Kingdom	• Statutory requirement to oblige water companies to prepare and maintain water resources management plans that look ahead 25 years.
European Commission	• Water Framework Directive, through groundwater quantitative status and surface environmental flows needed to ensure good ecological status.

Source: See country profiles associated with this publication on iLibrary as well as at *www.oecd.org/env/resources/waterandclimatechange.htm.*

Regulatory instruments to facilitate adaptation to extreme events have been identified by 21 countries as well as the European Commission. Most of the regulatory approaches identified to address extreme events aim to reduce flood risk through legislation and regulations on land-use and building and codes. For example, Germany has identified increased flood risks as one of its key vulnerabilities. Germany's physical planning law is under revision to adapt to the increased risk and to emphasise climate-friendly urban development as a guiding principle for the planning process.

Another example is the Netherlands' Room for the River programme. It was approved in 2007 in response to the recognition that extremely high river discharges from the Rhine tributaries will occur more frequently in the future. The measures aim to ensure that the rivers can discharge the greater volumes of water expected, without flooding. The extra room the rivers need to cope with higher discharges in the coming decades due to projected climate change will remain permanently available. Funding of EUR 2.1 billion has been set aside for a package of 39 measures for the period from 2008 to 2020. Several additional examples are captured in Table 3.4.

Regulatory measures have also been identified by a number of countries to cope with water quality issues (14 countries and the European Commission) and water supply and sanitation (13 countries and the European Commission). Many of the measures identified relate to existing regulations that deal with drinking water quality, wastewater treatment and water quality assessments. For instance, reviews of existing legislations on sewage disposal in view of climate change were undertaken to prevent additional deterioration of surface water quality.

Ten countries and the European Commission have identified regulatory instruments to address adaptation for ecosystems. Examples include legislation to protect ecosystems, environmental quality standards for protection of freshwater systems, measures to ensure minimum levels of groundwater and minimum ecological flows to contribute to maintaining good ecological status of water bodies.

Table 3.4. **Examples of regulatory measures to address extreme events**

Belgium	• Flood prevention plan in Brussels (*Plan Pluie*) promotes preventive measures to help ensure that the built environment is better adapted to increased precipitation. The flood prevention plan in Wallonia (PLUIES plan) aims to improve knowledge of the risk of flooding, reduce and decelerate run-off, improve river management and reduce vulnerability of flood-prone areas.
Czech Republic	• Prevention of Floods Programmes I and II support flood prevention measures to improve protection of the most endangered areas against floods.
France	• National Flash Flood Plan aims to ensure the safety of people facing flood hazards. • Integration of the impacts of climate change on natural risks into urban planning process.
Germany	• Revision of the physical planning law to emphasise climate-friendly urban development (climate protection and adaptation to climate change) as a guiding principle for the planning.
Japan	• Regulations and guidance on land use in disaster prone areas as well as unified flood control measures.
The Netherlands	• Room for the River: a package of 39 measures to ensure that the rivers can discharge the greater volumes of water expected, without flooding. • The Meuse project improves flood protection by increasing the peak discharge level that Meuse can handle by establishing a link between water and spatial planning. • Promotion of "climate buffers" to reduce the risk of flooding via temporary storage.
Slovak Republic	• The landscape revitalisation and integrated river basin management programme addresses flood prevention and protection.
Spain	• Review of the basic construction and design regulations and review of land planning and land uses to take into account increased climate risk due to climate change.
Sweden	• Amendment of the Planning and Building Act decrees that buildings may only be erected at suitable places and should take account risks of flooding and erosion in municipal plans.
Turkey	• Developing and ensuring the enforcement of the legislation on the structural effects of natural disasters caused by climate change.
United Kingdom	• Flood and Water Management Act to promote better and more comprehensive management of flood risks.
European Commission	• The Floods Directive requires that flood risk maps are drawn up by 2013 for river basins and associated coastal areas at risk of flooding. • Requires the establishment of Flood Risk Management Plans by 2015.

Source: See country profiles associated with this publication on iLibrary as well as at *www.oecd.org/env/resources/waterandclimatechange.htm*.

Economic instruments

Economic instruments influence behaviour via market signals rather than explicit regulation (so-called "command and control"). In an adaptation context, they can reduce baseline stress by promoting more efficient water use and allocation and provide a flexible means to adapt to changing conditions. Economic instruments also allow for decentralised decision-making about water use, at the level of the individual user. In contrast, a regulatory approach to water allocation typically means that all users (or groups of a particular types of users) are treated in the same way, which can dull incentives for inefficient users to improve water use practices. Relative to regulatory and information-based approaches, economic instruments are under-utilised in OECD countries' responses to adapt water systems. Table 3.5 depicts OECD countries' use of regulatory and/or economic instruments as a response to adapt to the various aspects of water systems.

In the OECD survey, 9 countries, along with the European Commission, identified economic instruments to cope with water quantity issues and 13 countries, along with the European Commission, have flagged their use to address extreme events, especially floods. In response to water quantity issues, countries have identified water pricing, abstraction charges, water-related taxes (an example is provided in Box 3.5), and water trading as part of their adaptation response. The use of water pricing and water trading to facilitate adaptation are discussed in more depth in Chapter 4. Table 3.6 provides some examples of economic instruments to address quantity issues.

Table 3.5. **Use of regulatory and economic instruments to adapt water systems**

	Regulatory Instruments					Economic Instruments				
	QN	QL	WSS	EX	ECO	QN	QL	WSS	EX	ECO
Australia	●				●	●			●	●
Austria										
Belgium	●	●	●	●		●		●	●	
Canada			Decentralised approach					Decentralised approach		
Chile	●	●	●		●					●
Czech Republic	●	●	●	●	●	●			●	
Denmark			●	●						
Estonia	●	●	●	●	●	●	●	●		●
Finland	●	●	●	●					●	
France	●	●	●	●	●	●		●	●	●
Germany	●	●	●	●		●	●	●		
Greece				●				●		
Hungary										
Iceland										
Ireland										
Israel								●		
Italy				●						
Japan				●						
Korea	●	●	●						●	
Luxembourg	●		●							
Mexico	●			●					●	
The Netherlands				●					●	
New Zealand	●	●	●	●						
Norway	●	●		●	●					
Poland		●		●	●					
Portugal										
Slovak Republic	●			●						
Slovenia						●	●	●		
Spain	●			●					●	
Sweden				●	●					
Switzerland	●	●			●				●	
Turkey	●	●		●	●	●	●	●	●	
United Kingdom	●		●	●		●			●	
United States		●	●	●					●	
European Commission	●	●	●	●	●	●	●	●	●	●

Note: QN = quantity; QL = quality; WSS = water supply and sanitation; EX = extreme events; ECO = ecosystems.
Source: See country profiles associated with this publication on iLibrary as well as at *www.oecd.org/env/resources/waterandclimatechange.htm.*

There are a few examples of countries, along with the European Commission, that have identified water pricing and water-related taxes as a part of their adaptation response. To address concerns related to water scarcity in Israel, tariffs in the domestic sector were raised by 40% in 2010 to promote more efficient water use. The EU's Water Framework Directive requires that the price charged to water users adequately covers costs, including environmental and resource costs, hence taking into account vulnerability to water scarcity and droughts. Slovenia has also identified water pricing measures to address excessive drinking water consumption and to promote sustainable and rational water use. Turkey is undertaking efforts to identify economic instruments to ensure efficient use of water.

Box 3.5. **France's "Rain Tax"**

France's "Rain Tax" was introduced to provide incentives for the improved management of urban rainwater. It was first introduced by the Law on Water and Aquatic Ecosystems (LEMA). Municipalities have the option of implementing the instrument if they choose to do so.

In France, municipalities are responsible for the management of urban rainwater, including collection, transport, storage, and treatment. A growing number of municipalities are confronted by the challenge of coping with increasing storm water runoff that strains the capacity of their current water treatment systems.

The new "Rain Tax" is designed to promote the sustainable management of rainwater, control pollution, prevent the risk of floods and also contribute to the financing of urban rainwater management. The tax applies to land and road owners (both public and private) in urban areas (and places where urbanisation is planned) according to the impermeable land surface (it will be imposed on surfaces of a minimum of 600 m^2, with the maximum tariff of EUR 1/m^2). However, if the land/road owner plans to create or improve its rainwater management system in order to limit or halt stormwater runoff, municipalities can offer a tax reduction of 20% to 100%. The revenue of the tax is exclusively dedicated to the public management of urban rainwater, such as the creation, operation, renewal and extension of infrastructure installation, and maintenance.

The Douaisis municipality was the first to put the new tax into place, starting in 2012.

Source: Ministère de l'Écologie, du Développement durable et de l'Énergie (MEDDE) (2006), *Taxe pour la gestion des eaux pluviales urbaines*, Fiche No. 3, MEDDE, Paris.

Table 3.6. **Examples of economic instruments to adapt to changing water quantity**

Australia	• Water trading to allow scarce water resources to be transferred to their most productive use and to dynamically respond to changing availability of and demand for water.
Belgium	• Flanders: ground water abstraction charge.
France	• Regular update of the water environmental taxes' rates in the Water Agencies' Intervention Programmes.
Slovenia	• Payments for water users' rights.
Turkey	• Identification of economic instruments to ensure the efficient use of water.
United Kingdom	• Promote greater trading of abstraction licences and bulk supplies of water to make supply system more flexible.
European Commission	• The WFD requires that the price charged to water users adequately covers costs, including environment and resource costs, hence taking into account vulnerability to water scarcity and droughts.

Source: See country profiles associated with this publication on iLibrary as well as at *www.oecd.org/env/resources/waterandclimatechange.htm*.

Water trading is also considered an adaptive response to changing water availability and demand for water. The UK is planning to encourage greater trading of abstraction licenses and bulk supplies of water to make supply systems more flexible. In Australia, water trading is used to allow scarce water resources to be transferred to their most productive uses. Water trading allows access to water resources to be reallocated over time in response to changing commodity prices, changing environmental conditions, changes to the size of cities and towns, and the changing availability of water. Water trading in Australia allows for the transfer of water access entitlements (permanent) and seasonal water allocations (temporary) between different entities. To improve the efficiency of the water market, the Australian Government has introduced water market charges and trading rules under the Water Act 2007. These rules improve the water market by freeing up and setting rules for trade and by ensuring appropriate price signals. In addition to water trading, best practice water pricing is a key element of Australia's National Water Initiative.

To address extreme events, the most frequently cited economic instrument is insurance to address flood risk (Table 3.7). In many countries, the central (or local) government plays a significant role in the insurance or reinsurance of flood risk. In all countries, the insurance sector is highly regulated. Climate change presents challenges to many existing flood insurance schemes, often requires modification of existing arrangements in order to reach efficiency and equity objectives, while remaining fiscally sound.

Table 3.7. **Examples of economic instruments to address extreme events**

Australia	• Government consultation on proposed reforms to flood insurance.
Belgium	• Recent reform of insurance for floods and other natural hazards.
Finland	• Reform of the compensation system for flood damages to better account for a changing climate and to respond to extreme weather conditions.
France	• Reform of the insurance scheme for natural catastrophes (CatNat).
Mexico	• Promotion of insurance as an aid to reduce vulnerability.
Spain	• Initiatives regarding insurance for natural disasters, including flood (e.g. adaptation of insurance market to better address climate change).
Switzerland	• Mandatory insurance schemes for natural disasters for all real estate owners.
Turkey	• Studies to be undertaken to explore the use of private and public insurance to address risks.
United Kingdom	• Statement of Principles between insurance companies and the Government commits insurance companies to continue to offer flood cover as part of standard policies (expired in June 2013 and an effort is underway to reach new agreement).
United States	• Reform of National Flood Insurance Programme (Biggert-Waters Flood Insurance Reform Act of 2012).
European Commission	• Potential implementation of payments for ecosystem services linked to natural water retention measures aimed at floods and drought prevention.

Source: See country profiles associated with this publication on iLibrary as well as at *www.oecd.org/env/resources/waterandclimatechange.htm*.

Several OECD countries are in the process of reforming existing insurance arrangements for flood risk or have recently adopted such reforms. The US recently passed the Biggert-Waters Flood Insurance Reform Act of 2012 to reform its National Flood Insurance Programme. France is currently undergoing a study to reform of the insurance scheme for natural catastrophes (CatNat). In the UK, a Statement of Principles (SoP) between the Government and the insurance industry was revised in 2008. The SoP commits insurance companies to continue to offer flood cover as part of standard policies. It will not be renewed following its expiry in June 2013. Therefore, an effort is underway by the Government to reach an agreement with insurers whereby insurance bills remain affordable without placing unsustainable costs on policyholders and the taxpayer. The examples of flood insurance in the UK, the US and France are discussed further in Chapter 4.

In case of Germany, a national insurance duty for natural hazards does not exist, and so there are awareness raising campaigns at the *Länder* level with the goal to raise the percentage of people and enterprises with voluntary insurance against natural hazards. Mexico is also currently promoting insurance in the context of adaptation.

Information-based instruments

By a large margin, information-based instruments are the most widely policy instruments by OECD countries to address adaptation for water. With the exception of Iceland, all OECD countries, along with the European Commission, have indicated their use of information-based instruments as a part of their adaptation response. The prevalence of the use of information-based instruments may reflect the important role of governments in overcoming information failures by building the evidence base to inform adaptation decisions (see Chapter 2 for further discussion). Governments are active in collecting, structuring, disseminating and sharing information on climate change adaptation and water risks. The use of information-based instruments is largely aimed at enhancing the scientific knowledge base, information sharing, and providing guidance for government actors at all levels as well as households and businesses. Table 3.8 shows the number of countries in the OECD survey that cite the use of information-based instruments as compared to other (regulatory and economic) instruments.

Table 3.8. **Frequency of use of regulatory, economic and information-based instruments cited in the OECD survey**

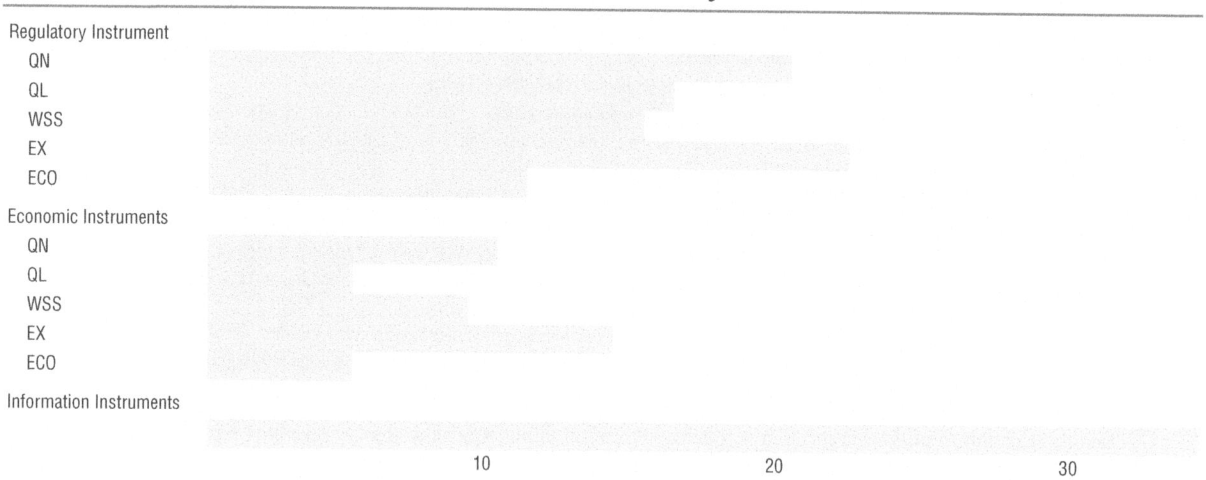

Note: QN = quantity; QL = quality; WSS = water supply and sanitation; EX = extreme events; ECO = ecosystems.
Source: See country profiles associated with this publication on iLibrary as well as at *www.oecd.org/env/resources/waterandclimatechange.htm*.

There are several types of information-based instruments. These include communication tools for raising awareness and capacity building as well as guidance documents for local governments, businesses or individuals. Information on risks often takes the form of early warning systems for hazards or flood risk maps.

Web-based platforms are one of the most frequently used mechanisms to disseminate information. Countries such as Austria, Czech Republic, France, Germany, and Norway and Switzerland have launched web-based information portals on climate change impacts and risks. Some of the web portals provide an expanded multi-tasking platform. Germany's KomPass is the national information, communication and co-operation platform on adaptation including a focus on water management. Norway's online portal "*Klimatilpasning Norge*" aims at strengthening the knowledge on climate adaptation and promotes exchange of information between sectors and levels of governments. This national web-based platform contributes to information on the effect of and consequences of climate change through specific advice and examples of climate change adaptation. At European level, European Commission has developed a web-based platform to facilitate exchange of information about climate change adaptation in European countries (Box 3.6).

Some of the web-portals focus specifically on water issues. Water specific information platforms include the Australian Water Resources Information System, the Canadian "Water and Climate Compendium", the Netherlands' "The Helpdesk Water", and the US's "WaterSMART". The Canadian "Water and Climate Compendium" is an online tool that provides access to information about climate change adaptation and water. The target

Box 3.6. European Climate Adaptation Platform: CLIMAT-ADAPT

The European Climate Adaptation Platform (CLIMATE-ADAPT) supports Europe in adapting to climate change. This platform provides a tool to help overcome a lack of information and sharing experience on climate change adaptation. CLIMAT-ADAPT helps users to access and share information on expected climate change in Europe, current and future vulnerability of regions and sectors, national and transnational adaptation strategies, adaptation case studies and potential adaptation options, and tools that support adaptation planning. CLIMATE-ADAPT organises information under the following main entry points:

- Adaptation information: Observations and scenarios, vulnerabilities and risks, adaptation measures, national adaptation strategies, research projects.
- EU sector policies: Agriculture and forestry, biodiversity, coastal areas, disaster risk reduction, financial, health, infrastructure, marine and fisheries, water management.
- Transnational regions, countries and urban areas.
- Tools: Adaptation support tool, case study search tool, map viewer.

Source: CLIMATE-ADAPT (2012), "European Climate Adaptation Platform", *http://climate-adapt.eea.europa.eu* (accessed 2 November 2012).

audience includes local/regional municipalities, rural communities, including indigenous communities, local watershed organisations, stewards, irrigation districts, private sector groups offering climate sensitive goods and services (e.g. agriculture, forestry, fisheries, utilities, mining, oil and gas, transportation, construction, manufacturing, professional services, recreation, real estate, finance and insurance), advocacy groups (e.g. non-government organisations), and provincial/territorial agencies with few resources.

A number of countries are providing adaptation guidance to local governments, businesses, and the general public. This has been a particular focus for countries taking a decentralised approach to adaptation. The Canadian Federal Government supported the development of a Guide for Municipal Climate Adaptation and a risk-based Guide for Local Governments. The Government of New Zealand provides technical manuals, summary publications, and guidance to inform local governments, businesses and individuals. Japan, Spain, Sweden, and UK also take a similar approach. Finland and Hungary provide a framework for co-operation and co-ordination among municipalities, research institutions, and public.

Table 3.9 provides examples of the range of information-based instruments used in response to flood risk. About half of the countries that identified floods or stormwater as primary concern have developed specific information-based instruments, such as risk maps and early warning services. France, Ireland, Luxembourg, Norway, Japan and the US have indicated the development of flood risk maps as part of their adaptation response. Belgium, Germany, Italy, Japan, Korea, Luxembourg, Turkey and the UK have developed flood-warning systems.

Examples of information-based responses to tackle the risk of shortage are less prevalent. Table 3.10 provides some examples. Japan and Luxembourg promote the use of rainwater to respond to drought and water scarcity. France has launched an information portal on drought and water restrictions. Belgium and Spain have initiated information campaigns to promote water savings. Portugal has a National Plan for Efficient Use of

Table 3.9. **Examples of information-based instruments to address the risk of flood**

Australia	• National guidelines for the flood risk information program.
Belgium	• Flanders: Public tool for flood prediction 48 h in advance.
Estonia	• Risk analysis of extreme events by cities (local plans take into account new flood risks).
Finland	• Flood forecasting and monitoring. • Stormwater management manual. • Pilot project on preparing for climate change in regional and general planning.
France	• Flood risk mapping for regions at high risk of flood. • Information portal on major risks, including natural disasters, such as floods.
Germany	• Platform on Flood Risk Information and Warning.
Greece	• Civil Protection planning and actions to address flood risk.
Ireland	• National Catchment-based Flood Risk Assessment and Management Programme. • Flood mapping website.
Italy	• Integrated early warning system for hydrologic risks.
Japan	• Regional disaster response and disaster prevention plans including hazard maps. • Information for flood warnings.
Korea	• Early warning systems to provide central and local governments.
Luxembourg	• Flood warning system. Flood risk maps and flood hazard maps.
The Netherlands	• "Living with water" programme (to stimulate co-operation between water management, spatial planning, science and practice, economy and sociology).
Norway	• The national mapping of flood and landslide risks.
Portugal	• Flood risk management plans.
Sweden	• Recommendations on flood risks in municipal planning.
Switzerland	• Guidelines and recommendations to identify risk areas and to implement water resource management and river basin management.
Turkey	• Risk maps for floods and landslides. • Developing monitoring early warning system for natural disaster. • Raising public awareness and developing capacities of relevant agencies on disaster response.
United Kingdom	• Flood warning service.
United States	• Flood risk mapping, assessment and planning. • Stormwater management.
European Commission	• Water Information System for Europe (online platform on water related information).

Source: See country profiles associated with this publication on iLibrary as well as at *www.oecd.org/env/resources/waterandclimatechange.htm*.

Water (PNUEA), approved in 2005, to promote the efficient use of water, particularly in the urban, agricultural and industrial sectors and to help minimise the risk of water shortages and improve environmental conditions in freshwater bodies.

A few countries have identified specific projects to address drought risk. In October 2012, Australian primary industries Ministers agreed to a new package of drought-related programs. The focus of this package will be on farmer risk management and preparedness, rather than government funded crisis support. In Italy, under the guidance of the National Committee to combat drought and desertification, six Local Action Plans were developed to adapt to increased drought and desertification.

Table 3.10. **Examples of information-based instruments to address the risk of shortage**

Australia	• Development of a new package of drought-related programmes.
Belgium	• Flanders: Information campaign to promote water savings.
France	• Information portal on drought and water restrictions (PROPLUVIA). • Identification of possible scenarios for adaptation of water-consuming activities in areas already suffering from water scarcity.
Italy	• National action plan to combat drought and desertification includes measures to address water and groundwater protection. • Pilot projects for adaptation to combat drought and desertification.
Japan	• Promotion of rainwater and reclaimed water use to respond to drought.
Luxembourg	• Promotion of water saving infrastructures and rainwater catchment system.
Portugal	• National Plan for Efficient Use of Water (PNUEA).
Slovenia	• Drought Management Center for South-Eastern Europe (promotes the development of tools for risk assessment of drought).
Spain	• Public campaign to promote water saving.
Turkey	• Preparation of guidance for water efficiency in industry and promoting pilot practices.
European Commission	• Drought Management Plans.

Source: See country profiles associated with this publication on iLibrary as well as at *www.oecd.org/env/resources/waterandclimatechange.htm*.

Financing approaches

Financing for adapting water systems has yet to be adequately addressed in most OECD countries. The adaptation strategies and plans of most OECD countries only briefly address financing issues, if at all. For the countries that have taken steps to addressing financing issues, several approaches have been taken (Figure 3.3). A few countries (e.g. Australia, Canada, France and Sweden) have allocated dedicated general adaptation funding from public budgets at the national level, some of which is allocated to water. Others (e.g. Germany and the UK) are mainstreaming adaptation actions into existing budgetary arrangements. Water-related support for adaptation is most often part of specific water programmes and projects (e.g. the Delta Fund in the Netherlands, Flood Prevention Programmes in the Czech Republic). A few OECD countries (e.g. Chile, Estonia, Hungary, Mexico, Slovenia, and Turkey) have received funding from international funding mechanisms (including EU Structural and Cohesion Funds) to advance adaptation of water systems. In addition, several countries (e.g. Germany, Denmark, France, Mexico, and the US) as well as the European Commission are exploring innovative financing mechanisms for adaptation (Box 4.4). In general, it is difficult to separate out financing for climate change adaptation-related expenditure from ordinary water expenditure.

A few countries have dedicated funding for general climate change adaptation from national budgets, some of which is allocated to water systems. France has allocated approximately EUR 171 million to implement measures in its National Adaptation Plan (2011-15). The Australian Government has invested over AUD 170 million over the period 2007-12 to fund national adaptation actions. This includes the establishment of the AUD 44 million CSIRO Climate Adaptation National Research Flagship and the AUD 126 million Climate Change Adaptation Program to lead Australian Government efforts to position Australia to manage risks of climate change impacts. In November 2011, the Canadian Minister of the Environment announced that the Government will spend CAD 149 million over the next 5 years on 10 adaptation programs from 9 departments and agencies (see Box 3.7).

Figure 3.3. **Various financing sources for climate change adaptation and water**

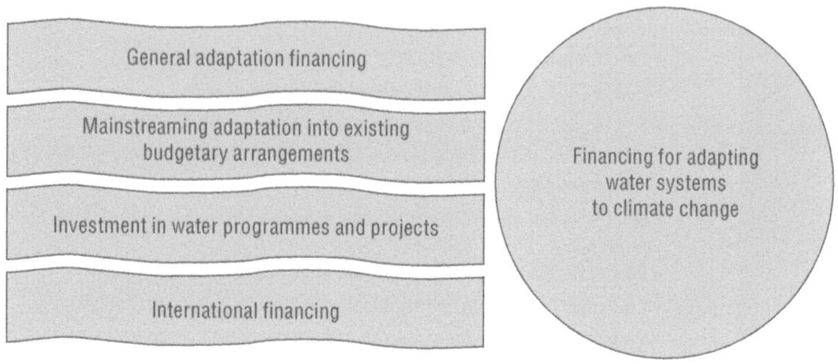

Source: See country profiles associated with this publication on iLibrary as well as at *www.oecd.org/env/resources/waterandclimatechange.htm.*

Box 3.7. **Canada's investment programmes for adaptation at the Federal and Regional Level**

The Federal Adaptation Policy Framework

The Framework was developed to support federal adaptation planning and highlights the importance of incorporating climate risk into decision-making. It sets the direction on national adaptation at a high level, recognising that the federal role must be limited and focused, and emphasising advancing scientific information, tools that underpin adaptation decision making, sharing knowledge, and assisting with the establishment of priorities for future actions. In November 2011, the Minister of the Environment announced the Government will spend CAD 148.8 million over the next 5 years on 10 adaptation programs from 9 departments and agencies to expand the number of departments and agencies involved and increases the overall level of federal funding for adaptation.

Regional Adaptation Collaboratives (RACs) of Canada

The Regional Adaptation Collaboratives (RACs) Climate Change Program, which ended in 2012, was a three-year, CAD 30 million, cost-shared federal program to help Canadians reduce the risks and maximise the opportunities posed by climate change. This Program helped communities prepare for and adapt to local impacts posed by the changing climate, such as decreasing fresh water supplies, increasing droughts, floods. Some examples are provided below.

- The British Columbia RAC, *Preparing for Climate Change: Securing B.C.'s Water Future,* focused on enhancing resilience to a changing climate and the anticipated related impacts on water and aquatic ecosystems.
- To address targeted climate change vulnerabilities, the Ontario RAC conducted a series of projects to advance decision-making in the areas of extreme weather risk management, water management and community development planning. Project activities will result in recommendations and actions for regulations, standards, best practices, plans, and/or polices, as well as tools to support adaptation action. Water management was a focus of the Québec RAC, which aimed to initiate adaptation on targeted issues by providing specialised information and tools to government water managers and watershed groups.

Source: See country profiles associated with this publication on iLibrary as well as at *www.oecd.org/env/resources/waterandclimatechange.htm.*

There are also cases where adaptation is mainstreamed into existing budgets. The German adaptation action plan will be funded from the budget of the respective governments departments within the current financial planning. England's adaptation programme has been designed on the basis that the funding will be mainstreamed into existing budgets. While funding has been provided for adaptation research, pressures on departmental resources are identified as one of the key challenges faced by the English adaptation programme (Mullan et al., 2013).

A significant number of countries have implemented or planned investments for specific water projects and programmes, which address climate change adaptation along with other drivers of water risk (e.g. socio-economic trends). The Australian Government is investing AUD 14 billion over 12 years to facilitate adaptation to climate change and as a response to increasing water scarcity via the *Water for the Future* programme (Box 3.8). In France, EUR 500 million will be invested over the period 2011-16 to implement the Drought Plan (*Plan sécheresse*) and the National Flood Plan (*le Plan submersions rapides*) and there are also several investment programmes for Water Agencies to reduce water loss and augment supply. These efforts aim at reducing water leakage within drinking water networks, supporting wastewater reuse and optimising water storage and development of new water storage, as needed. There are subsidies for environmentally-friendly water-related projects as well.

Box 3.8. **Australian Water for the Future Programme**

The Australian Government is investing AUD 14 billion over 12 years to facilitate adaptation to climate change and as a response to increasing water scarcity. The programme aims to improve water management and deliver a range of water policy reforms. Measures include:

- Sustainable Rural Water Use and Infrastructure Programme: The Australian Government has committed AUD 5.8 billion to increase water use efficiency in rural Australia. Under the programme, the Government has agreed to provide approximately AUD 3.2 billion for State Priority Projects in South Australia, New South Wales, Victoria, Queensland and the Australian Capital Territory.
- Restoring the Balance in the Murray-Darling Basin Programme: The Government has committed AUD 3.1 billion to facilitate water buy backs to protect and restore the environmental health of the Murray-Darling Basin. The programme allows irrigators to voluntarily sell their water entitlements to the Australian Government.
- Urban water programmes to secure and diversity urban water supplies: through the National Water Security Plan for Cities and Towns programme, the Australian Government has committed over AUD 250 million to fund practical projects that save water and reduce water losses in cities and towns. In addition, over AUD 680 million has been committed to investments in desalination plants, water recycling schemes, stormwater harvesting and reuse projects. Funding is also provided for research into improving the technologies and use of desalination and water recycling.
- Improving Water Information Programme: The Bureau of Meteorology received funding of AUD 450 million over ten years to revolutionise the way water information is measured, accounted for, reported, forecasted and analysed.

Source: See country profiles associated with this publication on iLibrary as well as at *www.oecd.org/env/resources/waterandclimatechange.htm*.

The Netherlands has established the Delta Fund, which covers the cost of measures and provisions for flood protection and freshwater supplies. Alongside construction and improvements, the Delta Fund also provides funding for the management, maintenance and operation of water works and related research. As of 2020, EUR 1 billion will be allocated by the Government for the implementation of the Delta Programme. The Delta Fund and the contribution from the water authorities ensure a secure source of financing for flood protection. The Netherlands also has a separate Flood Protection Programme that reinforces the primary defence structures along the coast, the rivers and the major delta waters that were shown not to meet the statutory standards. It has an available budget of EUR 2.5 billion for the 2009-20 periods, which will cover all measures to be taken.

In the UK, DEFRA expects to spend at least GBP 2.17 billion to address flooding and coastal erosion over the next four years 2012-15, an average of GBP 544 million per year. This sum consists of around GBP 1.04 billion in capital investments and around GBP 1.13 billion in other programme and administrative costs, such as maintenance, flood forecasting, and incident response.

Slovakia has a project to build a system of prevention flood-protection measures to reduce flood risk for 24 towns under the Landscape Revitalisation and Integrated River Basin Management programmes. It was financed through grants of approximately EUR 580 000 from the Prime Minister's budgetary reserves.

Investment programmes are also underway at the local level in a number of countries, funded by local government budgets or with support of central governments. For example, the German Environment Ministry introduced a funding scheme in 2011 to promote adaption to climate change at the level of individual enterprises and local authorities. This funding is expected to cover networking and education projects at the local/regional levels and to provide support for drawing up adaptation concepts. Funding is also available under the National Climate Initiative for municipalities for developing climate change adaptation concepts as part of municipal climate strategies. The Federal Government is funding exemplary models and demonstration schemes at local and regional level such as, Managing Climate Change in Regions for the Future (KLIMZUG), Urban strategies to Combat Climate Change, among others. This scheme provides financial incentives for frontrunners to foster innovation and to spread awareness about the necessity of adaptation.

In the UK, DEFRA remains committed to fully funding local authority new burdens under the Flood and Water Management Act. Up to GBP 36 million a year will be provided directly to lead local flood authorities. This investment is expected to deliver better flood protection to 145 000 households by March 2015.

A few OECD countries have relied on international funding mechanisms to support adaptation activities. For example, Chile has received support to develop activities related to climate change from the Global Environment Facility (GEF), its implementing agencies and bilateral development co-operation partners. In 2010, the World Bank approved a USD 450 million loan for Mexico to develop public policies aimed at supporting the Mexican government's efforts of promoting the adaptation of its water sector to climate change. The loan was intended to support government policies to contribute to the country's preparedness to confront the growing impacts of climate change through programs by the National Water Commission (CONAGUA).

The European Union provides some means to co-finance capital-intensive investment in water infrastructure and to help EU Member States comply with water legislation through EU Structural and Cohesion Funds. For the management of water resources, EUR 8 billion in total funding is provided for reducing leakage rates, connecting to water supply, generating additional supply and improving infrastructure. For disaster prevention, EUR 7 billion is available. The Solidarity Fund (EUSF) provides funds for disaster relief in member states. Around EUR 1 billion is allocated each year. Estonia's various activities concerning water management and climate change are financed in large part by the Cohesion Fund. In Slovenia, there are two ongoing flood defence projects receiving support from the Cohesion Fund. In Hungary, flood protection will continue to benefit from EU support in the frame of Environment and Energy Operational Programme, for which EUR 607 million has been allocated over the period of 2007-13.

Concluding remarks

While OECD countries are making progress on "knowing" the risk, much more could be done to build on current efforts to "target" and to "manage" water risks in response to climate change. The majority of efforts to date in OECD countries have focussed on "knowing" the risk by building the scientific evidence base and disseminating information. In terms of policy responses, information-based instruments are by far the most widely used. There are also numerous examples of using regulatory instruments to "target" and to "manage" water risks related to climate change. Yet, the OECD survey reveals that economic instruments are comparatively under-utilised thus far in the context of adaptation for water systems. While some of the literature points to the potential for using economic instruments, there remains a gap in terms of analysis and examples of how these instruments could be used to respond to specific adaptation challenges for water systems. Chapter 4 aims to help fill that gap by taking a closer look at how several economic instruments can facilitate efficient and timely adaptation for water systems and providing some illustrations from recent experience.

Notes

1. The individual country profiles are available at: *www.oecd.org/env/resources/waterandclimatechange.htm*.
2. The previous Assessment produced the report *Global Climate Change Impacts in the United States* in 2009, and the first National Assessment report was completed in 2000.
3. Examples of the use of various policy instruments for adaptation in Canada can be found at the sub-national level. For instance, several examples of the use of regulatory instruments in Ontario are provided in the country summary of Canada.

References

Australian Productivity Commission (2012), "Barriers to Effective Climate Change Adaptation", *Draft Report*, Austrian Government Productivity Commission, Canberra, *www.pc.gov.au/projects/inquiry/climate-change-adaptation* (accessed 11 November 2012).

CLIMATE-ADAPT (2012), "European Climate Adaptation Platform", *http://climate-adapt.eea.europa.eu* (accessed 2 November 2012).

ClimWatAdapt (2012), "Climate Adaptation: Modeling Water Scenarios and Sectoral Impacts", *www.climwatadapt.eu* (accessed 5 November 2012).

Department for Environment, Food and Rural Affairs (DEFRA) (2012), *UK Climate Change Risk Assessment: Government Report*, DEFRA, London, *www.gov.uk/government/publications/uk-climate-change-risk-assessment-government-report* (accessed 19 November 2012).

Ministère de l'Écologie, du Développement durable et de l'Énergie (MEDDE) (2011), *Eau et changement climatique : quelles stratégies d'adaptation possibles*, MEDDE, Paris.

Ministère de l'Écologie, du Développement durable et de l'Énergie (MEDDE) (2006), *Taxe pour la gestion des eaux pluviales urbaines*, Fiche No. 3, MEDDE, Paris.

Mullan, M. et al. (2013), "National Adaptation Planning: Lessons from OECD Countries", *OECD Environment Working Papers*, OECD Publishing, *http://dx.doi.org/10.1787/5k483jpfpsq1-en*.

The National Water Commission (NWC) (2011), *The National Water Initiative: Securing Australia's Water Future: 2011 Biennial Assessment*, NWC, Canberra, *http://nwc.gov.au/publications/topic/assessments/ba-2011* (accessed 18 November 2012).

The National Water Commission (2012), *Water Policy and Climate Change in Australia*, NWC, Canberra, *http://archive.nwc.gov.au/library/topic/markets/water-policy-and-climate-change-in-australia* (accessed 18 November 2012).

OECD (2012), *OECD Environmental Outlook to 2050: The Consequences of Inaction*, OECD Publishing, *http://dx.doi.org/10.1787/9789264122246-en*.

Chapter 4

Improving flexibility: Adaptive governance, policy options and financing approaches

An uncertain future for freshwater, the potentially rapid pace of change and the existence of possible irreversible tipping points increases the value of flexibility and calls for a dynamic, future-oriented approach to water governance and policy. This chapter highlights how "adaptive" water governance is gaining attention as a means to increase flexibility and deal with uncertainty related to long-term trends. It highlights how well-designed economic instruments can improve the efficiency and timeliness of adaptation responses by reducing baseline stress on water resources and providing flexibility to deal with increased variability, risks, and uncertainty. Based on a number of case studies it draws out lessons for adaptation on the use of insurance schemes, water trading, water pricing, and ecosystem-based approaches. Finally, the chapter examines some of the potential pitfalls in financing adaptation for water and looks at how a real options approach can be used to value flexibility in long-term investments.

Key messages

- **Adaptive water governance** and **sound water policy** will go a long way to enhancing resilience to climate change. At the same time, some existing water policies may also need to be adjusted to deal with increasing risk and uncertainty. Regulatory, information-based and economic instruments all have a significant role to play.
- Well-designed **economic instruments** can improve the efficiency and timeliness of adaptation responses by reducing baseline stress on water resources and hence, vulnerability. They can also provide flexibility to deal with increased variability, risks, and uncertainty and lower the cost of adjusting to changing conditions.
- **Flood insurance schemes** can provide incentives to reduce exposure and vulnerability to floods, efficiently spread residual risk, and offset the economic impact of floods. Greater uncertainty about the likelihood and severity of floods makes appropriately pricing flood insurance increasingly difficult.
- **Water trading** can allow efficient reallocation of water resources in response to changing conditions, including increasing variability and more frequent episodes of shortage. While temporary transfers can be effective for managing drought-induced supply variability, they are insufficient on their own to adjust to long-term changes in total water availability.
- Climate change strengthens the economic case for **efficient water pricing** that can reduce inefficient water use, encourage the diversification of sources of supply and raise financing for potentially higher investment needs. Prices could also be used to signal scarcity and hence the optimal timing for expanding supply. However, in practice, water has long been inefficiently priced in most cases and scarcity pricing has met with resistance.
- Incentives for **ecosystem-based adaptation** and **green infrastructure** can provide a cost-effective means to address uncertainty by avoiding or delaying lock-in to capital-intensive infrastructure, hence providing an additional "option" value. Although these approaches are gaining attention, especially in urban settings, experience to date remains preliminary.
- Adapting to climate change will likely add to the already substantial **financing gap** for water systems in OECD countries. It also raises several specific challenges due to long time frames and pervasive uncertainty. Attribution can also be an issue in the case of dedicated adaptation funding mechanisms. Financing adaptation should build on sound approaches to financing water systems generally, and avoid skewing financing to "speciality" projects that might be easily labelled as adaptation, but do not necessarily maximise net benefits.

It is well recognised that sound water policy will go a long way to enhancing resilience to climate change. Failure to adequately consider the policy context can constrain or undermine specific adaptation programmes or projects for water systems or result in mal-adaptation. Regulatory, information-based and economic instruments all have a role to play in effective and timely climate change adaptation. These instruments can be used in combination to "know", "target" and "manage" water risks to achieve an acceptable level of risk in a way that maximises social welfare over the long term. At the same time, some existing water and water-related policy settings may undermine effective and efficient adaptation, by distorting market signals or providing perverse incentives (e.g. subsidising water supply to certain users, encouraging development in areas at high risk of flood). These policies should be reviewed and adjusted in light of climate change and its impact on water security.

An appropriate policy mix employing a combination of regulatory, economic and information-based instruments is required to adequately address water risks. Table 4.1 provides examples from this policy toolkit. This chapter takes a closer look at how several economic instruments can facilitate more efficient and timely adaptation for water systems and provides some illustrations from recent experience. The specific focus on economic instruments was selected because these instruments tend to be under represented in both water policy and climate change adaptation policy discussions. As a result, they are often poorly understood or poorly applied in practice. Some literature on climate change adaptation for water points to the potential for using economic instruments, but there remains a gap in terms of analysis and examples of how these instruments could be used to respond to specific adaptation challenges. This chapter aims to help fill that gap. It also highlights some of the financing issues confronting adaptation for water systems and discusses some emerging approaches to address them.

Table 4.1. **Examples of water policy instruments to address water risks**

	Regulatory	Economic	Information-based
Risk of **water shortage** (including drought)	• Restriction on water use (e.g. hosepipe ban). • Administrative allocation of water. • Abstraction limits.	• Water pricing. • Water trading (e.g. water markets, water banks, dry year options). • Payments for ecosystem services (PES). • Microfinance schemes (e.g. to invest in rainwater tanks).	• Information and awareness campaigns to promote water saving. • Drought warning and information.
Risk of **inadequate quality**	• Water quality standards. • Pollution discharge permits.	• Pollution taxes, charges. • Tradable pollution permits. • PES.	• Information and awareness campaigns. • Technical assistance for improved farming techniques (to minimise negative impacts on water).
Risk of **excess** (including flood)	• Land use planning, zoning restrictions. • Building codes, standards.	• Flood insurance. • Public private partnerships (e.g. for flood defence structures). • PES.	• Flood risk maps. • Early warning systems.
Risk to the resilience of **freshwater systems**	• Minimum environmental flows.	• "Buy backs" of water entitlements from the water to ensure adequate environmental flows.	• Promoting awareness of the value of freshwater ecosystem services.

Adaptive water governance

Institutional frameworks and governance arrangements have an important influence on which policy instruments are the most appropriate in a given context and how they work in practice. Institutional fragmentation and poorly managed multi-level governance present obstacles to improved water management, especially in a context where regional, basin, and local authorities are usually in charge of water resources management and service delivery (OECD, 2011a). Effective public governance is a prerequisite for the effective implementation of policy instruments (e.g. abstraction limits, water pricing, and trading). It is also crucial to ensure sustainable financing and provide incentives for the efficient use of funds (OECD, 2011a).

While efforts to address climate change adaptation could provide an opportunity to revisit, and perhaps improve, existing governance arrangements, they will also likely strain existing multi-level governance challenges, both horizontal (across policy domains) and vertical (across levels of government). For example, while nationally led mandates may provide a strategic framework for adaptation in many OECD countries, climate change impacts on freshwater will be felt locally. Implementation of many adaptation responses will inevitably be local in nature, albeit conditioned by policy and institutional settings at all levels of government (supra-national, national, regional and local). Local development pathways shape exposure and vulnerability to water risks (e.g. land use planning affects flood risk). The broader policy environment may constrain or enable the efficient management of water risks at local level (e.g. cost-sharing arrangements across levels of government to provision flood protection or water storage). The way in which existing institutional arrangements affect the distribution of the costs and benefits of adaptation across levels of government and between the public and private sector will have a significant influence on what kind of adaptation occurs, where and when.

Co-ordination between climate and water policy communities is also important for effective adaptation. Addressing water resources as a priority theme or sector in the development of national or sub-national adaptation strategies and plans requires input and expertise from the water policy community. At the same time, mainstreaming climate change adaptation into water strategies, plans and policies requires integrating expertise and knowledge from the climate science community and may also benefit from guidance or tools developed for climate change adaptation more generally.

In addition to multi-level governance issues, climate change poses additional challenges to existing water governance arrangements. A non-stationary climate, the potentially rapid pace of change and the existence of possible irreversible tipping points increases the value of flexibility and calls for a dynamic, future-oriented approach that explicitly deals with uncertainty. Making the best use of constantly evolving scientific evidence characterised by significant uncertainty is a particular issue. As discussed in Chapter 1, there are significant gaps in the existing evidence base that pose challenges for informing practical site-specific adaptation decisions for water and may require new approaches and practices (see examples in Chapter 2). The science-policy interface is particularly important and there remains significant scope for improving the relevance and ease of use of climate science for practical adaptation decision making for water systems.

The long time frames involved in adaptation planning, along with the fact that measures incur upfront costs and have deferred benefits (in terms of avoided climate impacts, often difficult to quantify), also pose challenges for institutions' typical planning and policy cycles. There is also significant learning potential related to climate change adaptation, as the scientific evidence base improves, new approaches to adaptation are developed and practical experience is gained. Adequate mechanisms are needed to transmit new knowledge and feedback from experimentation at the local level to inform national level policy development.

Box 4.1. **Adaptive delta management in the Netherlands**

In the Netherlands, future socio-economic developments and changes in sea level, soil subsidence, river discharges and precipitation patterns are very uncertain. In response, the Delta Programme is using "adaptive delta management" to actively look for flexible strategies and highlight the added value of that flexibility. Adaptive delta management is not about deferring decisions or measures because of uncertainty, but rather about taking the right steps at the right time. It encourages an integral approach to tasking and reduces the risk of over- or underinvestment in future flood risk management and freshwater supplies.

Thinking long-term does not mean setting down measures today for the next 50-100 years. Solutions should be allowed to develop along with new insights and circumstances. However, it is advisable to guarantee that the solutions can be implemented in a cost-effective way when they are needed, and in the short term, to take the first steps that are worthwhile in every scenario, "no regret" measures.

Key points of adaptive delta management:

- Linking short-term decisions with long-term tasking.
- Incorporating flexibility in possible solution strategies (where effective).
- Working with multiple strategies that can be alternated between (e.g. adaptation pathways).
- Linking different investment agendas.

There are three key steps to implement this approach. First, it is important to clarify which short-term developments influence long-term tasking related to flood risk management and freshwater supply. Second, insight must be gained into the flexibility of the potential solutions for the tasking – e.g. is it easy to carry them out step by step and adjust them to accommodate actual developments? Finally, it is important to identify the decisions that are necessary in the short-term to enable the adaptive approach.

In developing the various "adaptation pathways", the circumstances under which it would be logical to move from one approach to another are studied along with how options can be kept open to actually enable that transition. The examination of adaptation pathways identifies both "no-regret" and "avoid-regret" (a measure has to be implemented in order to avoid a situation in which shifting to a different measure will no longer be possible or only at exorbitant cost) decisions for the short-term. Development paths are a powerful way to gain insight into which measure need to be taken when and how long-term tasking impacts short-term decisions.

The approach has already been applied in several of Delta sub-programmes. For example, the Rhine Estuary-Drechtsteden sub-programme developed adaptation pathways and explicitly identified when interventions would be required ("tipping points"), in the light of both flood risk management (e.g. dykes that no longer meet the standard) and freshwater supply (e.g. salinisation of intake points).

In order to take into account the added value of flexibility in the evaluation of Delta strategies in a systematic manner, the Netherlands Bureau for Economic and Policy Analysis is exploring options for a simple and broadly applicable method to structurally embed the added value of flexibility in economic analysis.

Source: Delta Programme, (2012), "Delta Programme 2013: Working on the Delta – the Road Towards the Delta Decisions", The Netherlands Ministry of Infrastructure and the Environment and Ministry of Economic Affairs, Agriculture and Innovation, *www.deltacommissaris.nl/english/Images/Delta_Programme_2013_ENG_tcm310-334162.pdf* (accessed 22 March 2013).

In response to the challenges posed by climate change and other drivers of water risk, "adaptive" water governance is gaining attention as a means to increase flexibility and deal with uncertainty related to long-term trends. The Delta Programme in the Netherlands has adopted an approach called "adaptive delta management" that consists of "phased decision-making that explicitly takes uncertain long-term developments into account in a transparent manner" (The Delta Programme, 2012) (Box 4.1). Experience with "adaptive" water governance is still preliminary, but is an important area that would be valuable to explore in greater depth in the future.

Existing political economy challenges to reforming water policies should not be underestimated as potential barriers to effective adaptation. Manifestations of past policies present significant obstacles to reform (e.g. historical land and water entitlements, existing infrastructures, stakeholder expectations). Crises may create political capital or windows of opportunity that can be used to enact water reforms, but they are not necessarily a precondition of reform (Winpenny, 2011). Aside from water crises, *per se*, water reformers can also take advantage of other types of crises (e.g. economic) and radical reforms (e.g. political transitions) to improve water policies. In the case of climate change adaptation for water, exceptional weather events that impose substantial costs on human lives and property may prove to be more catalytic than mounting scientific and economic evidence in terms of spurring action.

Improving incentives to manage risk and increasing flexibility in water policy

There are two key principles underlying the economic management of water – efficiency and equity (Grafton, forthcoming). Efficiency aims to maximise the welfare that is obtained from a resource by allocating it to its most valuable economic use. Equity concerns the distribution of resources across a given population. In the context of risk and uncertainty, adaptive efficiency is also important. Adaptive efficiency addresses the least cost path to maximise social welfare over the long term in the context of complex resources, unpredictability, feedback effects and path dependencies (Marshall, 2005).

Economic instruments can contribute to achieving the dual objectives of efficiency and equity. These are policy tools that influence behaviour through their impact on market signals rather than explicit regulation (Grafton, forthcoming). For example, water charges, pricing and trading can reduce baseline stress on water systems, building resilience to future climate change impacts by promoting efficiency in water use, allocating water to where it creates the most value and identifying low-cost options.

Economic instruments can also be used to achieve adaptive efficiency required for dynamic, decentralised and flexible responses to changing circumstances and deal with increased variability, risk and uncertainty. Water pricing and trading provide flexibility and help to minimise timing errors of adaptation actions by signalling scarcity and hence, the optimal timing for investments in supply augmentation. Adequate water pricing can encourage the development of alternative water supplies, providing supply diversification, thus improving reliability. Flood insurance schemes, properly designed, can provide incentives to reduce exposure and vulnerability to flood risks, spread residual risk and offset the economic impact of disasters. Incentives for ecosystem-based approaches and green infrastructures can provide cost-effective adaptation and provide flexibility in dealing with uncertainty by avoiding or delaying lock-in to more capital-intensive built infrastructures or costly retrofitting of existing infrastructures.

Many of these economic instruments are commonly used in water policy and are not specifically designed for adaptation. Previous work at the OECD and elsewhere has looked at the role these instruments play in promoting good water resources management, generally. This section examines how these instruments could be more systematically applied to facilitate climate change adaptation by not only reducing baseline stress on water systems, but also providing a flexible and cost-effective means to deal with increased variability, risk and uncertainty.

Flood insurance schemes

Insurance has long been used to deal with climate variability and weather risks. Population growth, the concentration of assets in exposed areas and climate change all contribute to the increasing costs of flood damage and create challenges for insurability (Swiss Re, 2012). Well-designed flood insurance schemes can provide incentives (through a price signal) to reduce exposure and vulnerability to risk, especially if premium discounts are awarded for risk reduction. Compensation in the event of a flood can offset the economic impact of the disaster and provide finance to restore damaged capital and speed up recovery. At the same time, flood insurance schemes need to avoid inadvertently promoting mal-adaptation, for example, by encourage development in high-risk areas and undermining incentives to adapt to long-term climate change. Designing insurance schemes that are priced to reflect actual risk while remaining affordable and offering comprehensive coverage will be increasingly challenging, and in some cases unviable, under climate change. Uncertainty about future flood risk will make efficiently pricing insurance increasingly difficult. A non-stationary future means that historical references will be an increasingly unreliable basis for the design of flood insurance.

Flood insurance schemes exist in various forms, including traditional indemnity-based insurance and index-based insurance. The type of insurance scheme and its design determines whether and to what extent it provides incentives for risk reduction and addresses problems such as moral hazard, asymmetric information, and adverse selection.[1] Traditional indemnity-based insurance covers the policyholder against the loss of an asset (a home or business). Although the design results in the payout being close to the actual loss incurred, there is a perverse incentive for the insured party not to undertake risk reduction if they know that the damage will be covered, hence creating moral hazard. Indemnity-based insurance also involves asymmetric information and may be prone to adverse selection. In addition, the process of settling claims can be time consuming and costly, thus entailing significant transaction costs (Agrawala and Fankhauser, 2008).

Index-based, or "parametric", insurance can address some of the problems related to indemnity-based insurance. This type of insurance makes a payment when a specific trigger event occurs, as opposed to indemnifying a specific loss. These insurance schemes may be more suitable for dealing with water risks related to climate change, as a number of weather conditions can be quantified and specified as a trigger event ahead of time. Parametric insurance can reduce moral hazard by decoupling the actual payout from the actual loss incurred, thus, preserving the incentive to reduce risk. However, this feature may also be a disadvantage, as actual payouts may not sufficiently compensate for losses. As there is no need for an assessment or verification of actual damage, the transaction costs are lowered and the speed of payout is improved. These features are particularly advantageous for dealing with catastrophic events (Agrawala and Fankhauser, 2008).

Flood insurance schemes can promote risk reduction in various ways. Insurers have effectively used differential premium pricing to discourage construction in high-risk areas (IPCC, 2012). Besides providing incentives for risk reduction via premiums, specific risk reduction measures can be required by insurance contracts. Insurance schemes also require a detailed analysis of risk, thus they can both raise awareness and provide valuable information (e.g. flood risk maps) to inform responses (ClimateWise, 2010). Insurers also typically monitor policyholders to ensure that loss-reducing measures required by contracts are actually implemented and adhered to (Botzen and van den Bergh, 2008). Insurers can also partner with governments and communities to establish appropriate regulatory frameworks and promote land use planning, building codes, emergency response and other policy responses to reduce flood risk (Botzen and van den Bergh, 2008; ClimateWise, 2010).

While well-designed flood insurance may facilitate adaptation, increasing flood risks due to climate change and other drivers present important challenges for flood insurance schemes. First, major weather events are occurring more frequently than in the past, which will mean higher expected losses and higher payouts, resulting in reduced time for insurers to recoup costs (Agrawala and Fankhauser, 2008; Thomas and Leichenko, 2011; IPCC, 2012). This trend makes it increasingly difficult to maintain affordability while pricing insurance efficiently (to reflect actual risk). This will limit the penetration of insurance coverage in cases where it is not compulsory. Subsidised premiums may increase the low uptake of insurance in certain areas, but also causes a shortfall between premium revenue and the payout of claims (Botzen and van den Bergh, 2008).

Uncertainty about future conditions also poses major obstacles for flood insurance schemes. It is becoming increasing difficult to price future flood risks, as historical references are a less reliable indicator of future trends. Despite some improvements in forecasting, a major challenge for the insurance sector is to improve the accuracy and resolution of hazard data and the likely impacts of climate change. As long as climate impacts are uncertain, insurance companies, which are risk-adverse themselves, will overcharge for climate risk and may refuse coverage of risks that might otherwise be insurable (Agrawala and Fankhauser, 2008). In addition, in most countries, the insurance industry is highly regulated, especially with regard to pricing of premiums, which limits the ability of insurers to adjust premium prices based on new evidence of climate change risks (Thomas and Leichenko, 2011). Overall, these challenges may restrict the availability of insurance and constrain its use as an instrument to facilitate adaptation.

Public policy measures may be needed to overcome some of these issues and facilitate sharing of flood risks between insurers and governments. For example, policy responses may take the form of publicly funded measures to bring risks (and hence premiums) down to an acceptable (and hence insurable) level (Agrawala and Fankhauser, 2008). This approach is reflected in the agreement by the UK Government and the insurance industry, called the "Statement of Principles", whereby insurers commit to continue to offer flood insurance to existing customers where they are at significant risk and the UK Environment Agency announced plans to reduce that risk within five years. A layered public-private system, where private insurers provide coverage up to a certain limit of damages followed by government provided insurance, has been proposed as an option for insuring against increased risks (Kunreuther, 2006; Litan, 2006; Botzen et al., 2009, *in* Thomas and Leinchenko, 2011). Broader use of premium subsidies, however, may reduce incentives to move away from activities that become progressively less viable under the changing climate (Skees et al., 2008, *in* Agrawala and Fankhauser, 2008).

Across the OECD, several different approaches to address flood damage exist, reflecting different risk sharing arrangements between the public and private sectors. Table 4.2 illustrates characteristics of these arrangements in the Netherlands, the United Kingdom, France and Germany. This section provides several examples of arrangements to address flood risk in the UK, the US, France and the Caribbean states and how they are being reviewed or reformed to address increasing risks from climate change and socio-economic drivers.

Table 4.2. **Arrangements against flood damage in the Netherlands, the UK, France and Germany**

Kind of arrangement	The Netherlands	The United Kingdom	France	Germany
Private coverage available	No	Yes	Yes	Yes
Premium differentiation	n.a.[1]	Yes	No	Yes
Public reinsurance	n.a.[1]	No	Yes	No
Public compensation scheme[2]	Yes	No	No[3]	Yes

1. Not applicable because private coverage is not generally available.
2. Does not involve a right to compensation.
3. Evidently, the public reinsurance scheme is (partly) financed through taxes.

Source: W.J.W. Botzen and J.C.J.M. van den Bergh (2008), "Insurance Against Climate Change and Flooding in the Netherlands: Present, Future, and Comparison with Other Countries", *Risk Analysis*, Vol. 28/2, Wiley-Blackwell, *http://dx.doi.org/10.1111/j.1539-6924.2008.01035.x*.

Case Study: Co-operation between public and private sectors to manage flood risk in the UK

The UK is one of very few countries that have a private market for flood risk insurance. Unlike many other countries, the UK government does not provide compensation in case of flood damage (Botzen and van den Bergh, 2008; ClimateWise, 2010). Over 5 million people in England and Wales live or work in properties that are at risk of flooding. In response to concerns about rising flood damages, the Association of British Insurers (ABI) and the UK government signed a voluntary agreement, called the "Statement of Principles" (SoP) in 2002, to ensure that flood risk is managed effectively and that competitively priced flood insurance remains widely available for households and small businesses. The agreement was most recently revised in 2008. Under the agreement, ABI members agreed to continue to make flood insurance available for households and small businesses as a feature of standard policies if the flood risk is not significant (e.g. no greater than 1 in 75 annual probability of flooding). Insurers also committed to continue to offer flood cover to existing customers at significant flood risk, provided that the Environment Agency announced plans to reduce the risk for those customers below significant levels within five years (HM Government, 2008).

Analysis by ClimateWise (2010) has identified several lessons from the UK experience. One of the key successes of the SoP has been in promoting a long-term strategy for flood risk management, taking into account the impact of climate change. It was also seen as a useful driver for an improved legislative framework for flood risk management in England and Wales, via the enactment of the Floods and Water Management Bill. The co-operation between insurers and the government has also shown to be effective in triggering collaboration at various levels. The various work-streams attached to the SoP on flood risk mapping, planning policy, investment strategy, property level resilience and access to insurance have led to collaboration among industry practitioners, civil servants and experts to improve flood risk management.

However, collaboration between the industry and government can lead to market distortions, which can have a negative effect on flood risk reduction efforts. For example, maintaining insurability despite significant risk exposure can undermine incentives for property owners to improve flood resistance of their properties. Also, the agreement that maintains current arrangements for properties at significant risk could hold back the development of specialist flood insurance more suitable for these properties. Finally, while risk-based pricing has been encouraged by both parties, its application in practice has proved difficult (ClimateWise, 2010). The SoP will come to an end on 30 June 2013 and the government and the ABI are continuing negotiations about risk sharing arrangements going forward. Clearly, a number of difficult issues will need to be addressed in terms of the balance between government's role in reducing flood risk and the role of insurance in transferring residual risk.

Case Study: Reforming the National Flood Insurance Programme in the US

The National Flood Insurance Program (NFIP) in the United States was created in 1968 to offer federally subsidised flood insurance for property owners and to promote land-use controls in floodplains (US Federal Emergency Management Agency, 2012). Participation in the programme is compulsory for properties with a federally-backed mortgage that are located in areas at risk of flooding at least once every 100 years. The programme's significant financial and operation challenges have been recognised for many years. Increasing risks due to climate change exacerbates these challenges. Concerns regarding the program's long-term financial solvency were heightened after unprecedented losses due to Hurricane Katrina in 2005.[2]

While the NFIP is intended to be fully-funded by premiums from policyholders, its design is not actuarially sound (US Government Accountability Office, 2010). A report from the Government Accountability Office (GAO) of the US Government (2010) pinpointed several design features that impeded the programme from more efficiently managing risk and constrained its ability to remain fiscally-sound, some of which were addressed in recent reforms. These features included statutory limits on rate increases and the inability to reject high-risk applicants. In addition, NFIP premiums did not reflect actual flood risk (nearly one in four property owners were paying subsidised rates) and the NFIP allowed "grandfathered" rates that permitted some property owners to continue paying rates that did not reflect reassessments of their property's flood risk. Further, the programme could not deny insurance on the basis of frequent losses, even though repetitive loss properties accounted for 25 to 30 per cent of claims, but only 1 per cent of policies (US GAO, 2010).

To address some of these challenges, the US Congress passed the Biggert-Waters Flood Insurance Reform Act in July 2012. The Act includes several reforms that could facilitate adaptation to flood impacts related to climate change. Key provisions of the recent reforms address the fiscal soundness of the programme, promote more efficient risk management and explicitly account for future changes to flood risk based on the best available scientific evidence. Analysis by Grannis (2012) highlights the Act's key provisions. These include the increase in premium rates of 20% annually (twice the previous limit) and the requirement that premiums be calculated based on "average historical loss year", including catastrophic loss years. Subsidies are phased out for a number of properties, in particular severe repetitive loss properties. To promote fiscal soundness, a Reserve Fund was created. The reforms also allow the Federal Emergency Management Agency (FEMA) to update flood insurance rate maps to include "future changes in sea levels, precipitation, and intensity of hurricanes", among other relevant information and data. The reform also extends flood insurance coverage at lower rates to communities that "have made adequate progress" in constructing or building flood control structures that protect from a 100-year flood (Grannis, 2012).

While these reforms are an important step forward in improving the efficiency and fiscal soundness of the programme, they are already being tested. The New York Times recently reported that early estimates suggest that Hurricane Sandy will rank as the nation's second-worst storm for claims paid out by the programme (Lipton et al., 2012). It is estimated that costs could reach USD 7 billion at a time when the programme is only allowed by law to add an additional USD 3 billion to its existing debt (Lipton et al., 2012).

Case Study: Assessing options to reform the "CatNat" scheme in France

In France, flood risk is addressed via a public-private partnership. Property insurance is not obligatory in France, although there is near universal coverage, with 99% of housing insured. Insurance for vehicles is obligatory. Under the "CatNat" scheme, coverage against flood risk and other natural hazards[3] is compulsory when the property is insured and included via a surcharge on property insurance provided by private insurers. The government sets a uniform rate for CatNat coverage (12% for a package policy for dwellings, 6% for an insurance contract for vehicles). This represents about EUR 1.3 billion per annum (Bommelaer et al., 2011). A portion of CatNat premiums is channelled into a state-managed fund for natural risk prevention, known as the Barnier Fund. Created in 1995, this fund was considerably reinforced recently, its resources growing from 2% to 12% of the CatNat premiums between 2007 and 2009. The estimated income of this fund in 2010 was EUR 154 million, of which more than EUR 140 million was allocated to flood prevention. Over the 1982-2006 period, 60% of the compensation paid for natural disasters (EUR 7.3 billion) concerned damage from floods (Bommelaer et al., 2011). The State also provides low-priced reinsurance with unlimited coverage via the Central Reinsurance Fund, guaranteed by State (Botzen and van den Bergh, 2008).

The CatNat scheme is based on the principle of solidarity in three main ways: *i)* the legal obligation for property insurance to provide cover for natural disasters; *ii)* all policy holders pay a uniform rate for the CatNat premium; and *iii)* the State guarantee to the Central Reinsurance Fund (Grislain-Letrémy and Peinturier, 2010; Bommelaer et al., 2011). Because coverage is mandatory, problems with adverse selection are reduced and nearly universal coverage is ensured. While insurance arrangements include deductibles to stimulate loss-reducing measures, the absence of differentiated premiums means that incentives to reduce risk are less than optimal (Botzen and van den Bergh, 2008).

Climate change is adding to the questions regarding the sustainability of the CatNat system and the effectiveness of measures to encourage risk reduction (Letremy and Grislain, 2009). Reforms of the CatNat scheme are currently undergoing study regarding the possible adjustment of insurance rates to support increased responsibility of individuals and businesses regarding their actual risk exposure. A law proposal to enable the adjustment of rates is currently under review in the Senate.

Case Study: Pooling catastrophe risk of excessive rainfall events in the Caribbean

The Caribbean Catastrophe Risk Insurance Facility (CCRIF) is the first and only multi-country parametric risk pool in the world. It is a regional catastrophe fund that provides coverage to Caribbean governments designed to limit the financial impact of disasters by quickly providing financial liquidity when a policy is triggered. It operates as a public-private partnership. The CCRIF was conceived in response to the severe damage caused by

Hurricane Ivan in 2004. This disaster caused billions of dollars of losses across the Caribbean, with losses close to 200% of GDP in both Grenada and the Cayman Islands. At the request of the Head of Governments of the Caribbean Community and with the assistance of the World Bank, the CCRIF was established to implement a cost-effective risk transfer programme for member governments (CCRIF, 2012a).

The CCRIF offers parametric insurance, which disburses funds based on the occurrence of a pre-defined level of hazard and impact, minimising delay and transaction costs imposed by an on-site assessment of losses. In May 2012, the CCRIF introduced a product for excessive rainfall. It is currently working with Swiss Re to generate a rainfall index, in order to inform the design of the policy. Data on exposure and vulnerability to excessive rainfall events generated by Swiss Re are used to produce rainfall risk profiles by the CCRIF. Premiums will be risk-based, so they will be determined as a function of the rainfall risk profile of each particular country and the coverage characteristics selected. Once rainfall risk profiles have been developed, the CCRIF will discuss coverage options with each country individually and policies can be issued once coverage levels have been agreed (CCRIF, 2012b).

Several features of the CCRIF products contribute to maintaining governments' incentives to invest in risk reduction. Premiums are based on estimates of countries' risk profiles, reflecting an analysis of actual risk. As a parametric scheme, potential compensation does not cover all potential damages, thus retaining incentives to undertake loss-reducing measures. Risk pooling offers the advantage of diversifying risk, hence greatly reducing the cost of reinsurance compared to the price each government would have paid individually (IPCC, 2012). By providing compensation quickly in response to a disaster, the human and economic costs of such disasters are reduced. This innovative approach to pooling catastrophe risk related to excessive rainfall events is promising and lessons from early experience should be useful to inform future adaptation decisions.

Water trading

Water trading is only one approach to allocating water resources and managing risk of scarcity (see Box 4.2 for an example of using operational guidelines to mitigate the consequences of extreme drought and address water availability in the Colorado Basin, US). However, water trading can promote efficiency in allocation and a flexible approach to

Box 4.2. **A co-ordinated approach to preserve flexibility to deal with water scarcity in the Colorado Basin, US**

Nearly 40 million people in the United States rely on the Colorado River for drinking water and populations that depend on the River are projected to increase to between 49 and 77 million by 2060 (USBOR, 2012). About 5.5 million acres of farmland are in production in the Basin. Climate models project that within this century, runoff in the Basin may be reduced by up to 20 per cent, due to reduced precipitation and temperature rise. By 2060, it is expected that commitments governing the allocation of Colorado River water (including the Colorado River Compact and the US Treaty with Mexico) will be met no more than 60 per cent of the time (USBOR, 2012). Water quality in the Colorado River may be affected by low soil-moisture conditions, predicted to be lower in the Southwest by 2050 than conditions experienced during any of the most severe droughts of the 21st century, including the 1930s Dust Bowl (Belnap and Campbell, 2011).

Box 4.2. **A co-ordinated approach to preserve flexibility to deal with water scarcity in the Colorado Basin, US** *(cont.)*

In 2007, the Colorado River Basin entered its eighth year of drought and the worst eight-year period in over 100 years of continuous recordkeeping. Storage in Colorado River reservoirs fell from approximately 94% of capacity in 1999 to a low of 52% capacity in 2004. A drought of this magnitude was the first of its kind in modern history for the Colorado River Basin and climate scientists suggest that droughts of this severity are likely to occur in the future. In May of 2005, the Department of the Interior began a public process to develop operational guidelines to mitigate consequences of extreme drought and address water availability in the lower basin during low-reservoir conditions.

After a two and a half year process of facilitating, analysing, and considering input from stakeholders including Governors' representatives of the seven Colorado River Basin States, a Record of Decision (ROD) was signed by the Secretary of the Interior in December 2007 to balance water supply, environmental protection, hydropower production, and recreation on the River. This ROD specifies interim guidelines that remain in effect through 2025. The guidelines are intended to provide contract users of Colorado River water certainty on the availability of water supplies during drought conditions. They include:

- Codification of Lead Mead elevations that define "normal", "surplus", and "shortage" conditions for deliveries to the Lower Basin States. The ROD defines how extra water during surplus conditions will be shared as well as how reduced deliveries during shortage conditions will be shared. These definitions are intended to "provide water users and managers in the Lower Basin with greater certainty to know when, and by how much, water deliveries will be reduced in drought and other low reservoir conditions".
- Establishment of four operational tiers based on water elevation in Lake Powell that trigger release amounts in the operational tiers from Lake Powell to Lake Mead in order to minimise shortages in the Lower Basin States and protect key reservoir elevations in Lake Powell. This addresses potential risk-risk tradeoffs with a "co-ordinated operation that would minimise shortages in the Lower Basin and avoid the risk of curtailments in the Upper Basin".
- Codification of rules for the creation, accounting, and delivery of Intentionally Created Surplus (ICS) to provide a "mechanism to encourage and account for augmentation and conservation of water supplies, referred to as ICS, that would minimise the likelihood and severity of potential future shortages".

This agreement represents an important evolution in the governance of the Colorado River, suggesting that the many interests in the basin can work together to address shared risks, concerns, and needs. The public process to develop ideas to address drought conditions in the Basin resulted in consensus among stakeholders to "encourage conservation, plan for shortages, implement closer co-ordination of operations of Lake Powell and Lake Mead, preserve flexibility to deal with further challenges such as climate change and deepening drought, implement operational rules for a long – *but not permanent* – period in order to gain valuable operating experience, and continue to have the federal government facilitate – *but not dictate* – informed decision-making in the Basin" (USBOR, 2007).

Source: Case study provided by the Arizona Water Science Centre, US, based on J. Belnap and D.H. Campbell (2011), "Effects of Climate Change and Land-use on Water Resources in the Upper Colorado River Basin: U.S. Geological Survey Fact Sheet 2010-3123", *http://pubs.usgs.gov/fs/2010/3123* (accessed 2 October 2012); United States Bureau of Reclamation (USBOR) (2012), "The Colorado River Basin Water Supply and Demand Study", *www.usbr.gov/lc/region/programs/crbstudy/finalreport/index.html* (accessed 14 March 2013); United States Bureau of Reclamation (USBOR) (2011), "Lake Powell Operations, Equalization and the Interim Guidelines", *www.usbr.gov/uc/rm/crsp/gc/Eq-IntGuide/Eq-IntGuidelines-Fact.pdf* (accessed 2 October 2012); United States Bureau of Reclamation (USBOR) (2007), "Record of Decision, Colorado River Interim Guidelines for Lower Basin Shortages and the Co-ordinated Operations for Lake Powell and Lake Mead: Final Environmental Impact Statement", *www.usbr.gov/lc/region/programs/strategies/RecordofDecision.pdf* (accessed 2 October 2012).

meeting future demand and dealing with uncertainty in the context of climate change. Variations of water trading arrangements include surface water markets, groundwater markets, water auctions, and water banks (Dinar et al., 1997). Water trading also allows access to water resources to be reallocated over time in response to changing conditions, including fluctuating commodity prices, changing environmental conditions, shifting demand for and availability of water. It promotes efficiency in allocation by allowing water transfers from areas of surplus to areas of scarcity and from low to higher value uses as well as creates incentives to use water efficiently.

The system of water rights which underlie water trading arrangements can also be used to more equitably share risks (Box 4.3). For example, this can be achieved by establishing rights in terms of proportional shares of an overall allocation as opposed to rights defined by a system of prior appropriation (e.g. first in time, first in line), which place

Box 4.3. **Water rights and risk sharing: Proportional rights vs. prior appropriation**

In the Murray-Darling Basin, Australia, risk of shortage is shared proportionally among water users. Australia's National Water Initiative set out two major principles regarding the sharing of risk arising from changes in the availability of water (Quiggin, 2011). The first principle established water allocations based on a share of available water, rather than a specific volume. Thus, in times of shortage, all users typically receive some water, but less than their full amount. However, during times of extreme shortage, there are circumstances where some entitlement holders might not receive any water. The second principle assigns risk arising from reductions in the overall availability of water for consumption depending on the reason for the change. Changes in water availability due to new knowledge about the hydrological capacity of the system will be borne by users. Reduction in water availability arising from changes in public policy, such as changes in environmental policy, will be borne by the public, which may imply compensation to users (Quiggin, 2011).

The system of water rights in many states in the Western US is based on prior appropriation that results in a continuum of senior right holders to junior rights holders. Appropriative rights are assigned in order of application of a quantity of water for a beneficial use. Those applications submitted earlier will be more senior to those submitted later ("first in time, first in line"). Water is then allocated according to seniority. In an extreme drought, even "senior" rights holders may not receive their allocation. In a mild drought, all but the most junior rights holders may receive full allocations. This system means that more junior users bear a greater risk of water shortage, while more senior users are relatively more insulated from risk. Compared to the system of prior appropriation, a system of water rights based on proportional shares, such as Australia's, allows for more flexibility in water use, provides incentives for all water users to take steps to conserve water and more equitable risk sharing.

Analysis of water resources in California under climate change by Hanemann et al. (2012) indicates that if the projections for sharp reductions in stream flows, increases in variability, and increased demand materialise, the current system of prior appropriation with seniority based on a historical hydrology may face growing political opposition. This may provide an opportunity to move to a new framework for water rights (with a grace period and perhaps some compensation), although any such changes in California would require extensive consideration and investigation. In the meantime, shoring up California's existing water rights systems will put California in a better position to adapt to

Box 4.3. **Water rights and risk sharing: Proportional rights vs. prior appropriation** *(cont.)*

climate change, by at least creating a baseline of use and supply, and by providing information about the pace of change in the water sector (Hanemann et al., 2012).

Source: Hanemann, M., D. Lambe and D. Farber (2012), "Climate Vulnerability and Adaptation Study for California: Legal Analysis of Barriers to Adaptation for California's Water Sector", *Public Interest Energy Research (PIER) Program White Paper*; Quiggin, J. (2011), "Managing Risk in the Murray-Darling Basin", in D. Connell and Q. Grafton (eds.), *Basin Futures: Water Reform in the Murray-Darling Basin*, Australia National University E-Press, pp. 313-326.

disproportionate risk of shortage on more junior rights holders. Markets for emerging derivative products for water, such as leases and forward contracts, may also provide more flexible arrangements to hedge risk, but experience to date is limited.

There are numerous requisites for the effective and efficient operation of water trading arrangements. Markets for water entitlements cannot alone resolve environmental, economic and social issues involved in the allocation of water across different uses (OECD, 2009). Well-defined and transferable property rights must exist, usually requiring the unbundling of land and water rights. The total number of rights must not be over allocated, taking into account environmental needs. The establishment and oversight of a properly functioning market requires an important role for governments to establish and adjudicate water rights, quantify, monitor and regulate harmful "third party effects", and provide the appropriate legal and institutional support (Dinar et al., 1997). Depending on the specific hydrological context, the relatively high cost of executing water transfers and lack of transport infrastructure may limit the scope for trade.

Case Study: Water markets in the Murray-Darling Basin, Australia

Australia's Murray-Darling Basin (MDB) is a well-known example of a comprehensive water market that has generated significant economic gains. The large majority of trading has occurred between irrigators – from low-value uses to higher-value uses. Despite severe reductions in water availability during the recent drought, it has been estimated that between 2006 and 2010 intra-regional trade and increased on-farm flexibility from water trading have provided benefits to irrigators of AUD 3.4 billion, as compared to scenarios without trade. Inter-regional trade over the same period contributed to an additional AUD 845 million in agricultural productivity (NWC, 2012).

Under the National Water Initiative (NWI), water trade allows the transfer of water access entitlements (permanent) and seasonal water allocations (temporary) between different entities. Water trading allows scarce water resources to be transferred to their most productive uses and allows access to water resources to be reallocated over time in response to changing conditions. The NWI also established means to equitably share risk of shortage among water users by establishing water allocations based on proportional shares of available water.[4] The Australian Government has introduced water market and charge rules and will introduce the trading rules in 2014 under the Water Act 2007 that will improve the water market by freeing up and setting rules for trade, and by ensuring appropriate price signals.

Extreme variability of inter-annual rainfall in areas of high population, agricultural and environmental significance is a key vulnerability for Australia in the context of climate change. Decreases in precipitation are expected across the country in the coming decades, with the largest decreases projected for central and southern Australia. Australia is also expected to experience more frequent droughts. A recent study by Jiang and Grafton (2012) looked at the role of water trading and the economic impacts of climate change and reduced surface water availability in the Murray-Darling Basin. It found that inter-regional water trade in periods of much reduced water availability reduces the negative on farm impacts of climate change. While the results show that losses to irrigated agriculture under a median climate change scenario are modest, under a "modified 2030 dry extreme scenario" there would be substantial reductions in water use, irrigated land use and profits. Nevertheless, the Basin-wide proportional economic impacts would be less than the percentage decline in water use. Thus, water trading, along with the development of drought-tolerant species and improved farming practices, could help irrigated agriculture adapt to climate change (Jiang and Grafton, 2012).

Case Study: Water banking and dry-year options in the Western US

In the Western US, experience has been gained in recent years with market-based instruments to manage risk associated with water scarcity and drought. Water banks and dry-year options are mechanisms that facilitate the voluntary, temporary water transfers during dry periods. Transfers may occur between different types of users – e.g. between agricultural users and cities or freshwater ecosystems – as well as the same type of users – e.g. from low-value to high-value crops. Such transfers have proved to be an essential means to transfer water to higher value uses and increase reliability for users that value it most highly (Colby and Pittenger, 2005). While the temporary nature of such transfers makes them effective for managing periodic scarcity, this makes them unsuitable on their own to provide reliable supplies over the long term and to adapt to long term changes in supply availability due to climate change impacts or increasing demand (Colby and Pittenger, 2005; Hanemann et al., 2012). In the case of California, while water marketing is playing an increasing role in coping with variability, long-term transactions (leases or permanent sales) are constrained by costs associated with environmental review and by the fact that many smaller users' water rights are essentially unquantified (Hanemann et al., 2012).

Colby and Pittenger (2005) define dry-year options as contracts that provide for temporary and voluntary water transfers in the event of drought. Buyers pay a fee to secure an option that will result in the transfer of water if the specified dry-year conditions are triggered. If the contract is triggered, buyers pay a set amount per acre-foot to exercise the option and receive the water transfer. While arid regions worldwide have experimented with dry-year option contracts, they have taken on an increasingly important role in the Western US in recent years. For example, in 2003, almost 100 000 acre-feet of water were transferred between the Metropolitan Water District of Southern California and Sacramento Valley irrigators via dry year options contracts. Because dry-year options are much more expensive than outright water purchases (often by a factor of four), the cost of using them to secure water supply need to be carefully weighed against the benefits provided in terms of reliability (Colby and Pittenger, 2005). While more expensive than permanent water transfers, the use of dry-year options are often employed to avoid third party impacts associated with permanent fallowing of agricultural land.

Water banks perform a range of functions to facilitate voluntary, temporary water transfers. Usually created to respond to drought conditions, a water bank is often used to facilitate the negotiation of temporary water transfers, in particular leases from irrigators. Water banks can also store water for future use, as is the case of the Arizona Water Banking

Authority, whose express purpose is to store water underground in aquifers in more abundant years for use in times of shortage (Megdal, 2007).

Dozens of regional water banks exist throughout the US. The Bureau of Reclamation created the Klamath Water Bank in 2003 to facilitate voluntary reductions in water diversions in order to ensure required flows for endangered fish populations (Colby and Pittenger, 2005). The Idaho Water Bank system traces its origins to the 1930s. The Bank facilitates the use of water rights to natural flow water or water stored in Idaho reservoirs. Water right holders can offer unused water rights to the Bank, which allows the water to be rented to other users (Idaho Water Resource Board, 2012).

The California Emergency Drought Water Bank (DWB) was established in 1991 as an adaptation mechanism to respond to one of the most severe droughts in the state's modern history. The DWB was created to buy water, mainly from agricultural users and water agencies in northern California, for resale to urban, municipal, and agricultural sectors in southern California. In the space of a few months, the DWR negotiated 351 contracts to purchase over 820 000 acre-feet of water. The offer price was set at USD 125 per acre-foot, and was resold for USD 175 per acre-foot to cover transaction costs of executing the transfer (Colby and Pittenger, 2005).

Analysis by Hanemann et al. (2012) points to several useful lessons from the experience of the DWB. While the transfers facilitated by the DWB were useful adaptations to deal with shortage, they were essentially temporary, and thus, exempt from meeting environmental requirements of the California Environmental Quality Act. There is no evidence that sellers would have been willing to transfer water for multiple years. Temporary responses may not be an adequate solution in the event that water shortages become more frequent under climate change. In addition, pumped groundwater was often used to substitute for surface water transferred to the Water Bank, reinforcing the tendency to overdraft groundwater that already exists. Transaction costs were high due to legal manoeuvring required to facilitate the transfers. Overall, reducing transactions costs and facilitating long-run transfers of water on a larger scale through the modification and better enforcement of surface water rights would be beneficial adaptive response to climate change (Hanemann et al., 2012).

Water pricing

Water pricing can promote water use efficiency and generate revenues to finance investments in water infrastructures and service provision. In general, putting the right price on water and water-related services encourages people to waste less, pollute less, and invest more in water services (OECD, 2012a). Increasing variability in rainfall, more frequent and severe droughts and greater uncertainty about future hydrological conditions due under climate change strengthen the economic case for efficient water pricing that can reduce inefficient water use, encourage the diversification of sources of supply and raise financing available for potentially higher investment needs.

Despite the good economic case for efficient water pricing that allows for sustainable cost recovery, most existing rate structures under price water. Water authorities often set prices without proper consideration of efficiency, which can lead to significant welfare losses (Grafton, *forthcoming*). In addition, most existing rate structures are inadequate as they are backward looking (relying on a historical cost basis) rather than forward looking (accounting for future replacement cost). As climate change, along with more stringent environmental and health standards, may increase the replacement cost of existing infrastructure, prices may need to increase to meet growing financing needs.

Increasing variability of water supply and uncertainty about future conditions in a changing climate also complicate the efficient timing of supply enhancements. While the costs of water supply augmentation are usually well-known, the inherent variability of water resources makes predicting the payback period for investment much more difficult (Hanemann, 2006; Grafton, forthcoming). Greater variability and uncertainty due to climate change will exacerbate this problem. For example, during a prolonged period of low rainfall, a large investment in water infrastructure may appear to be beneficial. However, a shift in available water supplies due to a break in drought conditions may make the supply augmentation unnecessary.

In theory, water pricing can also be used to effectively signal scarcity value of water and reduce demand during periods of scarcity. Scarcity pricing can signal the optimal time to invest in water infrastructure, so that supply is augmented efficiently (Grafton, *forthcoming*). Basically, scarcity prices work by triggering higher prices during periods of drought-induced excess demand. Higher prices make investments in water supply infrastructure more economically attractive, thus providing an incentive for the augmentation of water supply and evening out supply and demand for water. However, despite these theoretical arguments, scarcity pricing for water has not been put into practice to date. Another option to improve the efficiency and timing of investments in water supply infrastructures is the use of a real options approach to planning and investment, which is gaining interest in OECD countries.

In moving toward more efficient water pricing, ensuring the affordability of water services is also an important policy consideration. Water tariffs can be structured to account for the basic needs of all segments of the population (OECD, 2012a). Affordability for low-income households can be ensured, preferably through direct social transfer. Yet, challenges remain in order to gain social acceptability to raise water prices to efficient levels and to put scarcity pricing into practice.

While a full discussion of the complexities of water pricing is beyond the scope of this report, the following section provides some illustrations of how more efficient water pricing can facilitate climate change adaptation. For example, in some cases, it can promote efficient water use. Efficient pricing can also encourage the diversification of supply sources (e.g. recycled water or wastewater reuse), which builds the resilience of water systems to increased variability and prolonged periods of shortage. At the same time, systematically under-pricing water can encourage overuse and hold back investment in alternative sources of supply. However, diversification of supply sources is not only about pricing – regulatory barriers can also be significant, as well as issues of social acceptability. Examples from Israel, Australia and Spain provide insights for considering the role of water pricing in the context of climate change adaptation.

Case Study: Water pricing promotes efficient use and diversification of supply in Israel

Due to increasing water scarcity, water prices in the agricultural sector in Israel have risen by around 100% over the past decade. Price increases led to substantial changes in the use of agricultural water including: a move to drip irrigation, adoption of more appropriate crops, and an increase in the use of alternative water sources. As a result, agricultural water use has significantly decreased and saline and recycled sources of water now make up around 50% of irrigated water use. Despite the significant decline in agricultural water use, efficiency gains have meant that agricultural production has actually increased. Higher water prices and increased use of alternative sources of water have stimulated technological innovation and exports of water technology grew by around 20% year (OECD, 2010).

In the domestic sector, water tariffs were raised in 2010 by 40%, mainly to recover the cost of large-scale desalination plants.[5] Domestic users pay according to an increasing two-block tariff structure, which encourages water conservation. The addition of a third block with a much higher tariff that would apply to large water consumers in the event of exacerbated drought conditions was considered. The "drought tax" was initially applied in the summer of 2009 as a surcharge on water prices for consumption in excess of household allocations. This tax, however, was suspended in early 2010 in response to social protest and has not been reintroduced (OECD, 2011b).

Although scarcity prices have not yet been adopted in Israel, higher water prices, in combination with other measures (e.g. distinct level of security of volumes for alternative sources, versus less secure access to freshwater) have resulted in significant improvements in the efficiency of irrigated water use and the increased use of alternative sources of water. Higher water prices for the domestic sector encourage water conservation and allow for cost recovery of supply augmentation. The diversification of water supply and efficiency gains reduce Israel's vulnerability to increased variability of rainfall and more pronounced droughts. The experience of Israel also demonstrates the challenges of introducing scarcity pricing, even in countries where public awareness of water issues is high.

Case Study: Low water prices constrain the development of alternative sources, an example from Spain

In Spain, to meet the challenge of declining natural water availability and the limits to increasing the amount of abstracted "conventional" water resources, reused water and desalination have been playing an increasing role. Recycled water is used to supply public gardens, golf courses and selected irrigated agriculture as well as to recharge aquifers. The potential for further development of recycled water is relatively promising in Spain, in part due to the proximity of densely urbanised regions to intensive agriculture in dry regions. However, prices that adequately reflect costs are a condition for expanding the use of recycled water. The cost of producing recycled water often exceeds prices at current levels, slowing further development of this alternative source (Fuentes, 2011).

Spain is also relatively well-positioned to take advantage of desalination, especially along the dry Mediterranean coast, where pressures on water resources are particularly acute. Even so, production capacity in desalination is currently limited to a very small share of water supply. Despite the halving of production cost over the past ten years (according to government estimates), the cost of desalination still far exceeds the cost of conventional supplies and desalinated water is supplied at subsidised rates (Fuentes, 2011). Overall, the expansion of both conventional and unconventional water supply is constrained at current prices.

Case Study: Exploring options to improve the efficiency and timing of supply augmentation decisions, an example from Sydney, Australia

In 2007, Sydney, Australia commenced plans to build a desalination plant in response to concerns over water shortages. However, before the construction of the plant was completed, the drought ended, reducing pressure on water resources. To assess the welfare effects of investment in the desalination plant, Grafton and Ward (2010) evaluated the decision considering various combinations of volumetric price, water restrictions, and supply augmentation. The study found that the investment in desalination in Sydney was made prematurely, leading to welfare losses valued at hundreds of millions of dollars per year. These losses partly arose from the costs associated with using mandatory water restrictions and high volumetric water prices needed to cover the high capital costs associated with the premature construction of the desalination plant (Grafton and Ward, 2010). However, the study argues that losses could have been avoided if dynamically efficient volumetric pricing had been adopted in response to variability in water availability (Grafton and Ward, 2010).

Incentives for ecosystem-based approaches and green infrastructure

Ecosystem-based adaptation approaches involve making use of the services that biodiversity and ecosystems provide in order to adapt to the adverse effects of climate change (UNFCCC, 2011). Examples include restoring wetlands to reduce vulnerability to floods or improving catchment management to improve water quality or quantity. Green infrastructures use natural systems, such as vegetation and soil, to manage water. Ecosystem-based approaches and green infrastructure can be used in combination with or as an alternative to conventional "grey" infrastructures. As "no-regret" investments, often with multiple co-benefits (e.g. biodiversity), these approaches can be a cost-effective strategy to manage climate change impacts on water systems. They can also be effective strategies to address uncertainty, as these approaches are often less capital intensive and more easily reversible or adaptable than engineered alternatives, hence providing an additional "option" value. They can also provide a scalable complement to existing built infrastructure, allowing for incremental changes over time, as required.

Regulatory, economic and information-based policy instruments can be used to promote the use of ecosystem-based approaches and green infrastructure. Policy responses to encourage such approaches include tax incentives, land use planning, and payments for ecosystem services, among others. This section provides examples using several types of policy instruments.

In the context of climate change adaptation, ecosystem-based approaches and green infrastructure are gaining increasing attention.[6] While these approaches are not new, experience with using them in an adaptation context is just gaining ground. Challenges to implementing such schemes vary depending on the specific instruments used to put them in place. In general, putting these approaches in practice often requires a thorough understanding and assessment of the value of ecosystem services and adequate institutional capacity to establish, monitor, and enforce them.

Case Study: Recharging groundwater and managing stormwater with green spaces in Nagoya, Japan

Since the 1970s, the frequency of intense, localised rainfall events has increased in Nagoya, Japan's fourth largest city with 2.2 million inhabitants. Urbanisation has significantly encroached on green space in recent years, which has disrupted the natural water cycle. Surface sealing, for example, has decreased the volume of rainwater permeating into the ground. The amount of green space has been significantly reduced in the past decades, with green area making up only 25% of the city in 2005 (Yamada, 2010). The increased surface runoff of rainwater has increased pressure on existing sewer systems and rivers and increased urban flood damage. Evapotranspiration has been declining, which has exacerbated the heat island effect (Kamierczak and Carter, 2010; Yamada, 2010).

To address these challenges, the city of Nagoya has advanced efforts to promote the use of green infrastructures. Water and green corridors are used to promote flood control, cooling effects and ensure wildlife habitats (Yamada, 2010). The city's Water Cycle Revitalisation Plan (part of the Biodiversity Strategy) aims to increase the infiltration of water into the ground from the present level of 24% to 33% and to reduce runoff levels from 62% to 36% by 2050. This is to be achieved through protection and increased provision of green space, green roofs, permeable paving and structural measures (Kamierczak and Carter, 2010).

Nagoya is using a set of innovative incentives to promote the scheme: *i)* under the programme of preservation of existing green spaces, the City uses "loan for use" agreements with private green space landowners in order to secure favourable urban environments and provide the public with opportunities to experience local natural surroundings; *ii)* an incentive scheme for property developers, which allows them to increase the volume of their buildings if they reduce the total land footprint of the site and allow for the creation of continuous green areas (Kamierczak and Carter, 2010); and *iii)* in order to reduce the heat island effect and enhance water infiltration, the city recently established a requirement for tree planting on all plots of new development over 300 m^2, requiring greenery on 10-20% of the plot. The rule is now a prerequisite for planning permission (Commission for Architecture and the Built Environment, 2010).

Case Study: Reducing flood risks through the restoration of wetlands and green roofs in Denmark

Several cities in Denmark are using green infrastructure and ecosystem-based approaches to deal with heavy rainfall and increasing risk of flooding. Examples include using wetlands to reduce flood risks in Aarhus and an innovative approach to green roofs in Copenhagen (Danish Climate Change Adaptation portal, 2012).

In Denmark, more intense rainfall events and rising sea levels implied by climate change increase the urgency to provide flood protection for low-lying and densely populated areas. Using restored wetlands to hold water during and after extreme rainfall events and at high tide is viewed as an inexpensive solution to this challenge. A recently restored wetland, Egå Engsø, is being used to channel water from heavy rainfall, thus provisioning flood protection for the low-lying and densely populated area near Aarhus, Denmark's second largest city. The wetland also reduces nitrogen leaching from surrounding agriculture (Danish Climate Change Adaptation portal, 2012).

In 2007, following months of significant rainfall and a large amount of snowmelt, the limits of the flood protection systems were tested and required an emergency response to prevent a flood that could have had major economic consequences. This event highlighted the need for further preventive measures. A new wetland, Hede Enge, has been proposed to reduce risk from extreme rainfall events, which are projected to become more frequent and severe with climate change. The cost of the proposed project is estimated at approximately DKK 25 million, of which 80 per cent is for compensation for affected landowners for expropriation of land. Considered as a unique example of climate change adaptation, this project provides a good illustration of an ecosystem-based approach to adaptation for water systems (Danish Climate Change Adaptation portal, 2012).

In Copenhagen, an innovative green roof design has attracted international attention. The "8 House development", near the city centre, has two sloping green roofs that are exceptionally steep (30 degrees for one of the roofs, 32 degrees for the other) and wide (1 700 m^2) covered with drought-tolerant sedum plants. The building houses a day care centre, 476 flats, penthouses and townhouses, a café, businesses and shops, and has rapidly become a popular attraction for tourists as well as professionals interested to study the design on site. From a climate change adaptation perspective, the design provides effective stormwater management, mitigating the negative effects of heavy rainfall events. About 80% of the rain falling onto the surfaces evaporates, with the remaining water led directly into a flood retention basin. The roofs also combat the heat island effect (Danish Climate Change Adaptation portal, 2012).

Case Study: Managing stormwater with green roofs and "bluebelts" in New York City, US

In New York City, street, basement and sewer flooding is expected to become more frequent due to greater storm intensity and sea level rise due to climate change. Increasing stormwater and wastewater flows will be a challenge for the City's existing sewerage system. Built over hundreds of years, the current system is mostly gravity-based. The sunk-cost of the City's sewer systems is huge and there is almost no flexibility to modify existing piping, either in size or scope without extremely costly and disruptive retrofitting. Ecosystem-based and green infrastructure approaches have been identified as feasible and cost-effective alternatives (New York Department of Environmental Protection, 2008).

Several initiatives have been taken to promote the use of green infrastructure to manage stormwater, including expanding the use of natural landscape for drainage and run-off control, the modification of codes to increase the capture of stormwater and the provision of incentives for green infrastructure. Since 2007, USD 1.5 billion has been committed for green infrastructure to clean New York City waterways by making the city greener and more permeable. The initiative comes as part of Mayor Bloomberg's goal of making 90% of NYC's waterways suitable for recreation, which are currently being degraded by excess sewer and rain runoff. The City expects that this investment, combined with targeted cost-effective grey infrastructure, will reduce Combined Sewer Overflows (CSOs) by 40%. Compared with an "all-grey" approach, this plan is expected to save ratepayers more than USD 2 billion. In addition to improving the quality of the city's waterways, green infrastructure has a number of other benefits, including improvements in air quality, lower energy demand, reduce carbon emissions, increased species habitat and property values, and reduction in the city's vulnerability to the impacts of climate change (City of New York, 2011).

New York City is also using tax incentives to expand the use of green roofs by helping to offset their cost. Expansion of the "Bluebelt" programme, which provides runoff control using natural landscape, is promoting cost-effective stormwater management. The programme preserves natural drainage corridors, called "bluebelts", including streams, ponds and other wetlands areas, which allows them to perform their functions of conveying, storing and filtering stormwater, while providing community open spaces and diverse wildlife habitats. It is estimated that the Bluebelt programme saves tens of millions of dollars in infrastructure costs, when compared to providing conventional storm sewers for the same land area (New York City Department of Environmental Protection, 2008).

Financing issues: Avoiding potential pitfalls and accounting for option values

The cost of adapting to climate change will likely add to the already substantial financing gap for water systems in OECD countries. Investment needs in OECD countries are significant to confront the huge cost of modernising and upgrading their systems, so as to comply with increasingly stringent health and environmental regulations, maintain service quality over time, address pollution and growing populations, and in some cases, overcome years of neglect and under-financing. Estimates suggest that this could cost 0.35-1.2% of GDP a year over the next 20 years (OECD, 2012a).

Challenges for financing climate change adaptation for water

A range of factors will influence any additional cost imposed on water systems due to climate change adaptation. The nature and magnitude of specific climate impacts, the level of acceptable risk, and the timing of adaptation actions will have a significant influence on the cost of adaptation, and thus financing needs. In some cases, the additional cost imposed by adaptation needs may be marginal, relative to overall costs. This may be the case, for instance, for water quality, where other stressors may be the dominant cost drivers. In other cases, the additional cost for adaptation may be significant, such as in situations where natural water storage in the form of snowpack is destroyed due to rising temperatures and shifting precipitation regimes. Replacing this natural storage with infrastructure is likely to be very costly. In areas where precipitation is expected to increase, additional costs to manage floods may be significant (see Box 2.3 for a summary of evidence on costs of adaptation for water).

Societies' willingness to pay for adaptation will be influenced by its understanding of the risks faced and the level of risk considered acceptable. Cost-sharing arrangements between national governments and local communities will also influence the approach taken to manage water risks, and ultimately, the cost of doing so. For example, in cases where the cost of structural flood protection is partly or fully funded by national governments, while local communities bear the full opportunity cost of leaving flood plains undeveloped, incentives for local communities are skewed towards opting for structural approaches to manage flood risk, even if they may be more costly overall.

The timing of considering adaptation within the project cycle can also have an important bearing on costs, as well as the overall effectiveness of adaptation responses. Since water infrastructure projects have long lead times, if climate change adaptation is only considered towards the end of the process, (e.g. when financing is being sought) project developers may resist reconsidering the fundamental design and siting of the project in light of climate change considerations. This may result in a "bigger pipes" approach to adaptation, where safety margins are tacked onto projects that have already been conceived without consideration of climate change. A more effective and efficient approach to adaptation would take potential climate change impacts into account from the inception of the project and consider all possible risk management options, including possibly altering the design or siting of the project. If the projects account for adaptation at an early stage, it can be much cheaper than building add-ons or retro-fitting later on.

Beyond widening the financing gap, financing climate change adaptation raises several particular challenges for financing, due long time frames and pervasive uncertainty about future impacts. The expected cost of adaptation measures are usually known and incurred in the short term, while the expected benefits are more uncertain and accrue far into the future. This complicates the task of trying to determine an economically efficient level and timing of adaptation actions. Many water projects have very long asset lives (e.g. 80-100 years for dams), which means that taking climate change into account is essential to avoid mal-adaptation. However, project financing typically operates on a scale of 20 years or less, which may dull incentives for financiers (and also governments) to account for climate change impacts in the design of water projects.

Attribution issues may also pose a problem in the case of dedicated adaptation funding mechanisms. The problem of attribution arises from the fact that climate change adaptation typically occurs in the context of responding to a range of natural and socio-economic pressures on water systems. Adapting water systems to better deal with current climate variability also increases resilience to long-term climate change. Thus, identifying specific measures or actions that respond solely and exclusively to the impacts of long term climate change is both difficult and often impracticable. Indeed, mainstreaming climate change adaptation into water policies, programmes and projects is important to ensure that responses address a range of stressors and achieve overall water policy objectives at least cost.

However, mainstreaming can frustrate efforts to identify the "incremental" cost of adaptation. Attempts to identify this incremental cost are often driven by political imperatives motivating processes to account for how adaptation funding is spent. This is a particularly pressing issue for countries whose eligibility for climate finance is linked to demonstrating "additionality". While it is clearly important to promote accountability in the disbursement of dedicated adaptation funding and to ensure value for money, efforts to label financing for adaptation should avoid impeding mainstreaming and distorting the allocation of financing to "speciality" adaptation projects that may be easily labelled as "adaptation" but do not necessarily maximise net benefits.

Existing codes, standards and rules for economic valuation may also present a barrier to considering long time frames and dealing with uncertainty required for making decisions about adaptation investments for water. Discount rates appropriate for long frames (including declining rates) may be appropriate. The UK has addressed some of these issues with its supplementary guidance on "Accounting for the Effects of Climate Change" produced for the Treasury's "Green Book". The Green Book sets out the economic guidance used by the UK government to assess spending, investment and policy decisions. The supplementary guidance sets out the criteria that determine when it is particularly important to consider the risks and effects of climate change if a programme, policy or project. It provides tools for climate change risk assessment and offers real options analysis as an options appraisal framework, able to incorporate the uncertainty of climate change and the value of flexibility into decision making.

Mobilising financing for adapting water systems to climate change

Financing adaptation should build on sound approaches to financing water systems generally. For water supply and sanitation, this includes reducing costs (via efficiency gains or the choice of cheaper service options) and increasing the basic sources of finance – tariffs, taxes, and transfers (commonly known as the 3Ts) that can fill the financing gap. Repayable finance can be mobilised to bridge the financing gap, including from capital markets or from public sources. Improving the efficiency of operations can help to redress important losses of funds within the sector. Operational inefficiencies include poor revenue collection, distribution losses (leakage, or non-revenue water), labour inefficiencies and corruption. The choice of service levels and the choice of hardware and technologies to implement them can make a significant difference to costs (OECD, 2012a).

Financing for water resources management can rely on four principles. The first is the polluter pays principle, which can provide incentives to pollute less and/or generate revenues to alleviate pollution and compensate for welfare loss. The second is the beneficiary pays principle, which allows for the financial burden to be shared by those who benefit from water resources management. The third principle is equity to address affordability or competitiveness issues. Finally, coherence between policies that affect water resources (e.g. agriculture, land use, or energy policies) is the fourth principle. This principle recognises that factoring in the impacts on water into the allocation of public money to water-related sectors can be a more cost effective approach than mobilising additional funding for the water sector (OECD, 2012b).

As discussed in Chapter 3, experience with financing climate change adaptation for water systems in OECD countries is rather limited to date. The lack of economic evidence on the cost and benefits of adaptation for water may contribute to the relatively slow progress on financing issues. To shore up financing for adaptation, countries can further leverage existing sources of finance for water systems. To the extent that climate change adaptation is mainstreamed into water policies, programmes and investment decisions, existing sources of financing for water systems will also fund adaptation. In principle, the additional costs imposed by adaptation could be recouped via sustainable cost recovery. Yet, in cases where incremental costs will be substantial, this could make achieving sustainable cost recovery even more difficult.

In countries where dedicated funds for general climate change adaptation have been established, a portion of these funds is being channelled to water systems, given the priority that they are usually accorded in the context of national adaptation planning. Yet, water will continue to compete with other sectors for limited funds. A few OECD countries will continue to rely on international funding to support water and climate change adaptation efforts, including EU Structural and Cohesion Funds. Several countries are also exploring potential new sources of financing and innovative mechanisms (Box 4.4). To address pervasive uncertainty for water investments, real options approaches are gaining increasing attention. However, practical experience with the application of this approach for water investments in the context of adaptation is still limited.

Box 4.4. **Exploring innovative financing mechanisms for climate change adaptation and water**

Several OECD countries and the European Union are exploring innovative financing mechanisms to address climate change adaptation and water. Examples include:

- In **Denmark**, the government is currently scrutinising water sector legislation in order to prepare a new law proposal related to the financing of climate change adaptation of the water sector. The purpose of this work is to increase the possibilities for Danish water and sewer companies to finance more intelligent and socio-economic optimal climate change adaptation measures. For example, the proposal could make it possible for sewer companies to co-finance new measures on roads and in waterways, which keeps rainwater out of the sewer system.

Box 4.4. **Exploring innovative financing mechanisms for climate change adaptation and water** *(cont.)*

- The **European Commission** is considering the use of revenues generated from auctioning allowances under the Community greenhouse gas emission allowance trading system (EU ETS) for climate change adaptation. The EU White Paper *Adapting to Climate Change: Towards an European Framework for Action* (2009) supports the possibility of using such revenue for adaptation purposes. The revised Directive governing the EU ETS provides that at least 50% of the revenue generated from auctioning allowances should be used, *inter alia*, for adaptation in Member States and developing countries. The EU is also exploring the potential implementation of payments for ecosystems services linked to natural water retention measures aiming at the prevention of floods and droughts.
- The **German** Federal Government is examining the possibility of including aspects of climate adaptation in Federal funding programmes and joint funding instruments financed by the Federal Government, the *Länder* and the EU. The recent incorporation of adaptation into the funding instruments of the National Climate Protection Initiative is an example. The Federal Government also has a scheme to fund innovative initiatives and demonstration schemes at local and regional level. This scheme provides financial incentives to adaptation frontrunners to foster innovation and to spread awareness about the necessity of adaptation. At the end of 2011, the Environment Ministry (BMU) introduced a funding scheme promoting adaption to climate change at the level of individual enterprises and local authorities. This funding is expected to cover networking and education projects at the local/ regional levels and support for drawing up adaptation concepts.
- **Mexico's** *2030 Water Agenda* proposes to establish an Adaptation Contingency Fund that would improve Mexico's capacity to effectively replace or significantly modify water supply systems and flood systems. CONAGUA is still analysing alternatives for implementing the Fund. The recently adopted General Law for Climate Change specifies the need to create a fund for projects, studies, actions. In addition, since 2006, Mexico has been selling catastrophe bonds ("cat bonds") each year as an innovative form of risk financing. If a disaster occurs during a bond's lifetime, the government uses the money borrowed to pay for repairs. If no disaster occurs, the government pays the money back with interest. The latest such bond was issued on 15 October 2012 and raised USD 315 million.
- In the **United States**, legislation has been proposed that would establish an infrastructure bank to fund adaptation for water systems. The *Water Infrastructure Resiliency and Sustainability Act* was submitted to the US Congress in October 2011. The bill would authorise the Administrator of the EPA to establish a program of awarding grants to owners or operators of water systems to increase the resiliency or adaptability of the systems to any ongoing or forecasted changes to the hydrological conditions of a region of the US.
- As part of its National Adaptation Plan, **France** is undertaking a review of existing financing mechanisms to determine how they may be used in their existing form or potentially modified to support adaptation. It is also studying potential sources of additional financing.

Source: See country profiles associated with this publication on iLibrary as well as at *www.oecd.org/env/resources/waterandclimatechange.htm*.

Real options approaches

The pervasive uncertainty introduced by climate change poses challenges for considering investments in water infrastructure, which are typically capital intensive and long-lived, often with high sensitivity to climate. Low confidence in projections of future impacts presents a significant challenge to plan and design projects with potentially huge upfront capital investments to avoid risks that are difficult to quantify. When confronted with a range of possibly futures based on climate projections, there can be a temptation to look for an "average" future within that range, even if all of the projections may be equally likely and futures outside of the range of projections are also possible. There may also be a tendency to treat projections as predictions and overestimate the extent to which they can be relied on in a deterministic way. However, trying to build for an "average" future or misusing projections with low confidence as predictions can result in serious errors, including significant mismatches between water infrastructures and future climate that may require costly retrofitting or result in stranded assets. Instead, using investment approaches that accommodate uncertainty and value flexibility may be required.[7]

Real options approaches have been gaining increasing attention by OECD governments in the context of climate change adaptation. For example, the UK's Green Book supplementary guidance on adaptation offers real options analysis as an options appraisal framework to incorporate the uncertainty of climate change and the value of flexibility into decision-making. A real options approach has been employed to assess water-related projects under climate change in The Netherlands. In its report on urban water, the Australian Productivity Commission proposed wider use of real options approaches for water supply augmentation decisions. The Commission's modelling indicated that applying a real options approach could reduce the cost of supply for two cities in Australia (Melbourne and Perth), by over AUD 1 billion over a 10 year period, compared with traditional approaches to planning and investment.[8] Given the advantages in dealing with risk and uncertainty, as compared to traditional planning and investment approaches, the National Water Commission and the Water Services Association of Australia have endorsed the real options approach to planning and investment (Government of Australia Productivity Commission, 2011).

Real options analysis explicitly incorporates the value of flexibility into decision-making. A "real option" is an alternative that can be put into place, adjusted or discarded as new information is gained. It is particularly useful in cases where sunk costs are high, projects are scalable and have long lead times and there is an expectation of learning over time. In the case of climate change, water investments could be readjusted to respond to higher or lower magnitude impacts, sooner or later than expected as knowledge about future conditions improves.

Similar to a standard costs benefits analysis, a quantitative real options appraisal compares discounted costs and benefits over time to generate a net present value, but incorporates an additional step in order to account for the value of flexibility. A decision tree can be used to map out the sequence of project actions, decision (or "trigger") points and key events. Information on costs, benefits and probabilities associated with different options can be used to calculate how payoffs change according to different future scenarios. (HM Treasury, 2009). An illustration is provided in Box 4.5.

Case Study: Real options approaches for the Thames Estuary 2100 Project, UK

The Environment Agency's Thames Estuary 2100 (TE2100) project is developing a strategy for tidal flood risk management to the year 2100. The Thames estuary floodplain contains 1.25 million people (one sixth of London's population), about GBP 200 billion of property, and key transport and infrastructure assets, including the London Underground, 16 hospitals and eight power stations. Given the value of assets at risk, the long lead times involved in developing solutions and the uncertainty of future climate effects and the

Box 4.5. **Illustration of using a real options approach for flood protection**

A simple decision tree can be used to set out two options for flood protection – an investment in a fixed levee today or the investment in a portion of an upgradable levee today, with an option to upgrade the levee in the future. Each option is evaluated under two possible climate change scenarios (severe or moderate climate change).

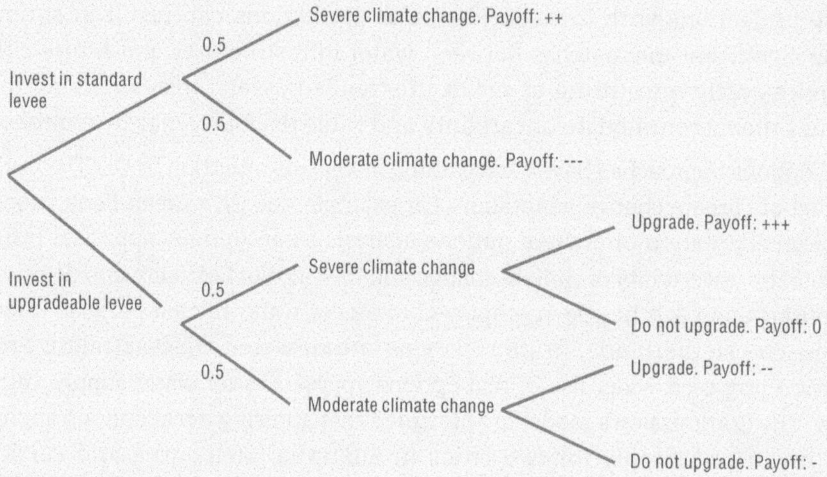

In this example, two future climate scenarios, severe or moderate, are equally likely scenarios (probability of 0.50). The high flood wall is built today and has positive net benefits only under the severe climate scenario. Under the moderate scenario, the levee has a negative payoff.

In the case of the upgradeable wall, the first portion is built today, with an option upgrade in the future. Under the severe climate change scenario, a higher payoff can be attained than in the case of the standard levee because the upgrade can be scaled as appropriate and a portion of the total cost of the wall deferred into the future. Thus, in terms of discounted net present value, the upgradable wall under the severe climate scenario has the superior payoff. In the case of moderate change, the most appropriate option is to not upgrade the wall, as additional flood protection is not needed, minimising potential losses associated with the initial investment.

Source: Adapted from HM Treasury (2009), "Supplementary Green Book Guidance: Accounting for the Effects of Climate Change". *http://archive.defra.gov.uk/environment/climate/documents/adaptation-guidance.pdf* (accessed 11 November 2012).

potential for learning, a flexible, adaptive approach to incorporating climate change has been taken. The project identified options to cope with different levels of sea level rise, and the thresholds at which they will be required. The options were designed to implement the small incremental changes common to all options first, leaving major irreversible decisions as far as possible into the future. The strategy can be reappraised in light of the new information and options can be brought forward (or put back) (HM Treasury, 2009).

One issue in using a real options approach is the possibility of some options being prematurely closed off or ruled out, for example through the actions of private property owners. For TE2100, one example of this is the potential development of areas which may be needed for future flood risk management activities (new defences, flood storage areas, etc.). As a result, the TE2100 strategy is likely to recommend the safeguarding of land in order to keep these options open. This entails the opportunity cost of foregoing development on the land, but provides an option value. If it is possible to minimise the opportunity cost of not developing relatively small parcels of land (for example by making other sites available through the land use planning system), then the value of maintaining options for protecting London from increasing flood risk is arguably large (HM Treasury, 2009).

Concluding remarks

Given the scale of the challenge, governments need to explore the full range of options improve the flexibility of water governance, policy and financing approaches. "Adaptive" water governance is gaining attention as a means to increase flexibility and deal with uncertainty related to long term trends. Drawing lessons from early experience with adaptive governance approaches will be important for steering adaptation responses in the future.

In terms of policy responses, regulatory, information-based and economic instruments all have a role to play to "know", "target" and "manage" water risks. Most water policy instruments were not specifically designed with climate change adaptation in mind and may need to be adjusted in light of new evidence to better address increasing risk and uncertainty. At the same time, climate change strengthens the case for addressing existing inefficiencies in current settings.

To date, economic instruments are comparatively under-explored in the context of water policy and climate change adaptation. Economic instruments are just one part of the policy toolkit, but a potentially powerful one, when properly designed and carefully implemented. In the context of increasing variability and declining predictability, they can provide flexibility and minimise the costs of adjusting to changing conditions. Climate change provides opportunities for more systematic use of these instruments, but also challenges.

Finally, the long time frames and pervasive uncertainty about future climate change impacts pose challenges for financing adaptation for water. Climate change is also likely to contribute to existing funding shortfalls. In cases where dedicated adaptation financing is available, an excessive emphasis on additionality can undermine effectiveness by skewing funding towards projects that may be more easily labelled as "adaptation", but do not necessarily maximise net benefits. Sound financing approaches are called for along with the use of flexible investment strategies where appropriate.

Notes

1. Moral hazard occurs when the person making decisions involving risk taking does not bear the full cost of potential negative consequences. Asymmetric information exists when one party in an exchange has better information than another. In the context of insurance schemes, insurers will not have perfect information about the risk taking behaviour of the insured, leading to under- or over-estimation of risk and hence, a certain amount of inefficiency. Adverse selection is the propensity of persons with higher risk to buy insurance more frequently and in greater amounts as compared to those with lower risk. This situation may come about where there is asymmetric information and insurers are unable to reflect this effect in the price of insurance.
2. As of August 2010, NFIP's debt to the US Treasury stood at USD 18.8 billion (Government Accountability Office, 2010).
3. With the exception of storms and hail.
4. While there is some variation among States, generally, water entitlements are divided into categories by the level of risk that an entitlement holder is willing to accept. Higher security

entitlements are met with available water before allocations are made to general or low security entitlements.

5. While the expansion of desalination reduces reliance of more variable conventional supplies, the high energy intensity of the production process means that it may be considered "mal-adaptive" in the context of climate change, due to their contribution to increasing greenhouse gas emissions. However, the energy source used (e.g. fossil fuel, renewable, etc.) and the degree of energy efficiency bear considerably on the actual level of greenhouse gas emissions. Desalination plants in Israel are among the most energy efficient and cost efficient in the world.

6. See the UNFCCC *Nairobi Work Programme's Database* on ecosystem-based approaches to adaptation, *http://unfccc.int/adaptation/nairobi_work_programme/knowledge_resources_and_publications/items/6227.php*; The Green and Blue Space Adaptation for Urban Areas and Eco Towns (GRaBS) project website, *www.grabs-eu.org*.

7. See Hallegate et al. (2012), for a recent review and an assessment of approaches for making investments under deep uncertainty.

8. See Government of Australia Productivity Commission (2011), "Australia's Urban Water Sector", *Productivity Commission Inquiry Report*, Vol. 2, No. 55, 31 August.

References

Agrawala, S. and S. Fankhauser (eds.) (2008), *Economic Aspects of Adaptation to Climate Change: Costs, Benefits and Policy Instruments*, OECD Publishing, *http://dx.doi.org/10.1787/9789264046214-en*.

Belnap, J., and D.H. Campbell (2011), "Effects of Climate Change and Land-Use on Water Resources in the Upper Colorado River Basin: US Geological Survey Fact Sheet 2010-3123", *http://pubs.usgs.gov/ fs/2010/ 3123* (accessed 2 October 2012).

Bommelaer, O., J. Devaux, C. Noel (2011), "Financing Water Resources Management in France: A Case Study for an OECD Report", *Collection Études et documents*, Department for the Economics, Assessment and Integration of Sustainable Development, Department of the Commission-General for Sustainable Development.

Botzen, W.J.W. and J.C.J.M. van den Bergh (2008), "Insurance Against Climate Change and Flooding in the Netherlands: Present, Future, and Comparison with Other Countries", *Risk Analysis*, Vol. 28/2, Wiley-Blackwell, *http://dx.doi.org/10.1111/j.1539-6924.2008.01035.x*.

Caribbean Catastrophe Risk Insurance Facility (CCRIF) (2012a), "Understanding CCRIF's Hurricane and Earthquake Policies", *Technical Paper Series #1*, *www.ccrif.org/sites/default/files/publications/TechnicalPaper1-HurricaneEarthquakePoliciesAugust2012.pdf* (accessed 18 November 2012).

CCRIF (2012b), "Newsletter", Vol. 3/3.

City of New York (2011), "PlaNYC Update: A Greener, Greater, New York", *http://nytelecom.vo. llnwd.net/o15/ agencies/planyc2030/pdf/planyc_2011_planyc_full_report.pdf* (accessed 15 November 2012).

ClimateWise (2010), "Adapting to the Extreme Weather Impacts of Climate Change: How can the Insurance Industry Help?", *www.rms.com/Publications/ClimateWise_Adaptation_Report_120910.pdf* (accessed 5 November 2012).

Colby, B.G. and K. Pittenger (2005), "Structuring Voluntary Dry Year Transfers", New Mexico Water Resources Research Institute, pp. 149-154.

Commission for Architecture and the Built Environment (2010), "Developing an Urban Heat Island Strategy", Greater London Authority, *http://webarchive.nationalarchives.gov.uk/20110118095356, www.cabe.org.uk/sustainable-places/advice/urban-heat-island-strategy* (accessed 15 November 2012).

Danish Portal for Climate Change Adaptation (2012), *www.klimatilpasning.dk/en-US/Sider/ClimateChange Adaptation.aspx* (accessed 14 November, 2012).

Delta Programme, (2012), "Delta Programme 2013: Working on the Delta – the Road Towards the Delta Decisions", The Netherlands Ministry of Infrastructure and the Environment and Ministry of Economic Affairs, Agriculture and Innovation, *www.deltacommissaris.nl/english/Images/ Delta_Programme_2013_ENG_tcm310-334162.pdf* (accessed 22 March 2013).

Dinar, A., M.W. Rosegrant and R. Meinzen-Dick (1997), "Water Allocation Mechanisms: Principles and Examples", *Policy Research Working Paper*, No. WPS 1779, The World Bank, Washington, DC, *http:// documents.worldbank.org/curated/en/1997/06/694756/water-allocation-mechanisms-principles-examples* (accessed 10 October 2012).

Fuentes, A. (2011), "Policies Towards a Sustainable Use of Water in Spain", *OECD Economics Department Working Papers*, No. 840, OECD Publishing, Paris, *http://dx.doi.org/10.1787/5kgj3l0ggczt-en*.

Government of Australia Productivity Commission (2011), "Australia's Urban Water Sector", Productivity Commission Inquiry Report, Vol. 2/55.

Grislain-Letrémy, C. and C. Peinturier, (2010), "Le Régime d'assurance des catastrophes naturelles en France métropolitaine entre 1995 et 2006", *Collection Études et documents*, Department for the Economics, Assessment and Integration of Sustainable Development, Department of the Commission-General for Sustainable Development.

Grafton, R.Q. (*forthcoming*), "Economic Instruments for Water Management", *OECD Environment Working Papers*, OECD Publishing, Paris.

Grafton, R.Q. and M.B. Ward (2010), "Dynamically Efficient Urban Water Policy", *Centre for Water Economics, Environment and Policy Research Paper*, pp. 10-13, Australia National University, Canberra.

Grannis, J. (2012), "Analysis of How the Flood Insurance Reform Act of 2012 (H.R. 4348) May Affect State and Local Adaptation Efforts", *Georgetown Climate Centre, www.georgetownclimate.org/analysis-of-the-flood-insurance-reauthorization-and-reform-law-2012* (accessed 15 December 2012).

Hanemann, M. (2006), "An Economic Conception of Water", *in* P.P. Rogers, M.R. Llamas and L. Martinez-Cortina (eds.), *Water Crisis: Myth or Reality?*, Taylor & Francis plc., London.

Hanemann, M., D. Lambe, and D. Farber (2012), "Climate Vulnerability and Adaptation Study for California: Legal Analysis of Barriers to Adaptation for California's Water Sector", *Public Interest Energy Research (PIER) Program White Paper.*

HM Government (2008), "Revised Statement of Principles on the Provision of Flood Insurance", DEFRA, *http://archive.defra.gov.uk/environment/flooding/documents/interim2/sop-insurance-agreement-080709.pdf* (accessed 5 November 2012).

HM Treasury (2009), "Supplementary Green Book Guidance: Accounting for the Effects of Climate Change", *http://archive.defra.gov.uk/environment/climate/documents/adaptation-guidance.pdf* (accessed 11 November 2012).

Idaho Water Resource Board (2012), "Water Supply Bank", *www.idwr.idaho.gov/WaterManagement/WaterRights/waterSupply/ws_default.htm* (accessed 14 December 2012).

Intergovernmental Panel for Climate Change (IPCC) (2012), "Managing the Risks of Extreme Events and Disasters to Advance Climate Change Adaptation", *A Special Report of Working Groups I and II* of the Intergovernmental Panel on Climate Change, Cambridge University Press, Cambridge, UK and New York, NY, USA.

Jiang, Q. and R.Q. Grafton (2012), "Economic Effects of Climate Change in the Murray-Darling Basin, Australia", *Agricultural Systems*, Vol. 110, Elsevier Science, pp. 10-16. *http://dx.doi.org/10.1016/j.agsy.2012.03.009*.

Kazmierczak, A. and J. Carter (2010), "Adaptation to Climate Change Using Green and Blue Infrastructure", *Database of case studies, Green and Blue Space Adaptation for Urban Areas and Eco Towns* (GRaBS) project website, *www.grabs-eu.org* (accessed 15 November 2012).

Kunreuther, H.C. (2006), "Has the Time Come for Comprehensive Natural Disaster Insurance?", *in* R.J. Daniels, D.F. Kettl and H. Kunreuther (eds.), *On Risk and Disaster: Lessons fromHurricane Katrina*, University of Pennsylvania Press, Philadelphia, PA.

Lipton, E., F. Barringer and M.W. Walsh (2012), "Flood Insurance, Already Fragile, Faces New Stress", *The New York Times, www.nytimes.com/2012/11/13/nyregion/federal-flood-insurance-program-faces-new-stress.html?pagewanted=all&_r=0* (accessed 7 December 2012).

Litan, R. (2006), *Sharing and Reducing the Financial Risks of Future "Mega-catastrophes"*, The Brookings Institution, Washington, DC.

Marshall, G.R. (2005), *Economics for Collaborative Environmental Management: Regenerating the Commons*, Earthscan, London.

Megdal, S.B. (2007), "Arizona's Recharge and Recovery Programs", *in* B. Colby and K. Jacobs (eds.) *Arizona Water Policy: Management Innovations in an Urbanizing, Arid Region*, RFF Press.

National Water Commission (2012), "The Impacts of Water Trading in the Southern Murray-Darling Basin Between 2006-07 and 2010-11", National Water Commission, Canberra.

New York City Department of Environmental Protection (NYCDEP) (2008), "Climate Change Program Assessment and Action Plan", *Report 1*, *www.nyc.gov/html/dep/html/news/climate_change_report_05-08.shtml* (accessed 14 November 2012).

OECD (2013), "Water Security for Better Lives", *OECD Studies on Water*, OECD Publishing, *http://dx.doi.org/10.1787/9789264202405-en*.

OECD (2012a), *Meeting the Water Reform Challenge*, OECD Publishing, *http://dx.doi.org/10.1787/9789264170001-en*.

OECD (2012b), *A Framework for Financing Water Resources Management*, OECD Publishing, *http://dx.doi.org/10.1787/22245081*.

OECD (2011a), *Water Governance in OECD Countries: A Multi-level Approach*, OECD Publishing, *http://dx.doi.org/10.1787/9789264119284-en*.

OECD (2011b), *Environmental Performance Review: Israel 2011*, OECD Publishing, *http://dx.doi.org/ 10.1787/9789264117563-en*.

OECD (2010), *Taxation, Innovation and the Environment*, OECD Publishing, *http://dx.doi.org/10.1787/9789264087637-en*.

OECD (2009), *Managing Water for All: An OECD Perspective on Pricing and Financing*, OECD Publishing, *http://dx.doi.org/10.1787/9789264059498-en*.

Quiggin, J. (2011), "Managing Risk in the Murray-Darling Basin", *in* D. Connell, and Q. Grafton (eds.), *Basin Futures: Water Reform in the Murray-Darling Basin*, Australia National University E-Press, pp. 313-326.

Skees, J.R., B.J. Barnett and B. Collier (2008), "Agricultural Insurance – Background and Context for Climate Adaptation Discussions", *Consultant Report to the OECD*.

Swiss Re (2012), "Flood: An Underestimated Risk", *http://media.swissre.com/documents/Flood.pdf* (accessed 20 November 2012).

Thomas, A. and R. Leichenko (2011), "Adaptation through Insurance: Lessons from the National Flood Insurance Programme", *International Journal of Climate Change Strategies and Management*, Vol. 3/3, pp. 250-263.

United Nations Framework Convention on Climate Change (UNFCCC) (2011), "Ecosystem-Based Approaches to Adaptation: Compilation of Information", United Nations Office, Geneva.

United States Bureau of Reclamation (USBOR) (2012), "The Colorado River Basin Water Supply and Demand Study", *www.usbr.gov/lc/region/programs/crbstudy/finalreport/index.html* (accessed 14 March 2013).

United States Bureau of Reclamation (USBOR) (2011), "Lake Powell Operations, Equalization and the Interim Guidelines", *www.usbr.gov/uc/rm/crsp/gc/Eq-IntGuide/Eq-IntGuidelines-Fact.pdf* (accessed 2 October 2012).

United States Bureau of Reclamation (USBOR) (2007), "Record of Decision, Colorado River Interim Guidelines for Lower Basin Shortages and the Co-ordinated Operations for Lake Powell and Lake Mead: Final Environmental Impact Statement", *www.usbr.gov/lc/region/programs/strategies/RecordofDecision.pdf* (accessed 2 October 2012).

United States Federal Emergency Management Agency (FEMA) (2012), "The National Flood Insurance Program", *www.fema.gov/national-flood-insurance-program* (accessed 11 November 2012).

United States Government (2012), "Biggert-Waters Flood Insurance Reform Act of 2012", in *Moving Ahead for Progress in the 21st Century Act*, United States Government, Washington, DC.

United States Government Accountability Office (GAO) (2010), "National Flood Insurance Program Continued Actions Needed to Address Financial and Operational Issues, Testimony Before the Committee on Banking, Housing and Urban Affairs, US Senate, Statement of Orice Williams Brown, Director Financial Markets and Community Investments", GAO, *www.gao.gov/highrisk/risks/ insurance/national_flood_insurance.php* (accessed 3 August 2012).

Winpenny, J. (2011), "Report of the Rapporteur", *OECD Global Forum on Environment: Making Water Reform Happen*, *www.oecd.org/environment/biodiversitywaterandnaturalresourcemanagement/49769095.pdf* (accessed 4 October 2012).

Yamada, M. (2010), "Soil, Water, Green: Toward Biodiversity Nagoya", *www.kankyo-net.city.nagoya.jp/citysummit2010/images/kaigi/session2/04_nagoya.pdf* (accessed 14 November 2012).

Chapter 5

Using hindsight to guide the future: Concluding remarks

The scientific evidence documenting the range and significance of climate change impacts on freshwater resources is compelling and growing. Given the importance of water as an essential resource as well as a potential threat, climate change impacts on freshwater will affect not only water and flood management *per se*, but also a number of key policy domains (e.g. energy, agriculture, infrastructure, biodiversity, and health). Despite the ever-expanding scientific basis, reliable information about the nature, magnitude and timing of impacts at the scale needed for practical, site-specific adaptation is generally lacking. Adaptation decisions need to accommodate significant uncertainty. Effective and timely adaptation can lower the cost of climate change impacts.

In both the near and longer term, adaptation to climate change calls for a flexible, dynamic, and future-oriented approach that takes into account climate variability on all timescales along with associated changes in human and natural systems. In the absence of accurate and precise climate predictions, a risk-based approach can explicitly accommodate a range of possible futures to inform adaptation decision-making by identifying priorities and options to manage risk and enhance resilience at least cost to society.

Governments should seek to "know", "target" and "manage" water risks – risk of shortage (including drought), risk of excess (including floods), risk of inadequate quality, and risk of disrupting the resilience of freshwater ecosystems. While climate change adaptation should be considered in the broader context of water security and not in an isolated manner, it is also critical aspect of, even a prerequisite for, achieving resilient water security over the long-term. Delaying or neglecting to systematically integrate climate change into water management policies, planning and practice can threaten water security and make its achievement more costly over time.

While there has been significant discussion and debate regarding what governments *could do* or *should do* to adapt water systems to climate change, a systematic review of what governments *are doing* was needed to gauge progress and gain insights from practical adaptation efforts. To address this gap, the OECD Secretariat undertook a survey of policies for water and climate change adaptation across all 34 member countries and the European Commission. The results provide an overview of general trends and lessons learned, sketching a broad picture adaptation in practice.

The survey demonstrates that all OECD countries have observed changes to freshwater systems and nearly all anticipate increasing water risks in a changing climate. While OECD countries are making progress on "knowing" the risk, much more could be done to build on current efforts to "target" and to "manage" water risks. The majority of efforts to date in OECD countries have focussed on building the scientific evidence base and disseminating information. Given the significant investments being made to improve the scientific evidence base, there is an urgent need to ensure that this evidence is used to best effect and meets the needs of users making practical, on-site adaptation decisions.

In terms of policy responses, information-based instruments are by far the most widely used. There is significant potential to scale up the use of economic instruments (e.g. insurance schemes, water pricing and water trading, among others) as well as ecosystem-based approaches and green infrastructure. Economic instruments are just one part of the policy toolkit, but a potentially powerful one, when properly designed and carefully implemented. In the context of increasing variability and declining predictability, they can provide flexibility and minimise the costs of adjusting to changing conditions. Climate change provides opportunities for more systematic use of these instruments, but also challenges.

Even though climate change is likely to add to the existing funding shortfall for water, most countries have yet to explicitly address financing issues for adaptation. Governments should avoid the potential pitfalls that can come with over-emphasising the "additionality" of adaptation projects at the expense of sound approaches that seek to maximise net benefits of investments. Flexible investment strategies can also help to address the challenges of long time frames and pervasive uncertainty.

Climate change is a novel challenge that will test conventional approaches. However, valuable lessons can be drawn from existing experience in dealing with climate variability and water disasters. History is full of examples of water crises and examples of how they could have been avoided or better managed. Indeed, hindsight can provide valuable lessons for climate change adaptation and help to prepare for the future.

The case for action is well-documented and compelling. Yet, experience shows that mounting scientific evidence and a robust economic case do not necessarily spur action. Not all water risks can be avoided, but well-prepared, resilient water systems will be better able to adjust to new conditions, at lower cost, and bounce back from disasters more quickly. Reforming water systems takes time, stakeholder engagement, and political will, so there is an urgent need to get started. Countries that take steps to shore up their water governance today and put in place the policies needed to prepare for the future can avoid managing water crises in a reactive and more costly way.

Country Profiles

Australia

Climate change impacts on water systems

Observed changes and trends	• Annual average daily mean temperature has increased by 0.9 °C from 1910 to 2011. • During recent decades, there has been a general trend towards increased spring and summer monsoonal rainfall across Australia's north, higher than normal rainfall across the centre, and decreased late autumn and winter rainfall across the south. • Reductions in rainfall over southern Australia, coupled with increases in temperatures, have led to dramatic reductions in stream flow in these regions. • Increase in the intensity and frequency of extreme temperatures and severe drought.				
Projected impacts	• Increase in temperature of 0.5 °C to 1 °C by 2030 and of 0.5 °C to 2 °C by 2070 above 1990 levels under the "optimistic" scenario. • Increase in temperature of 1 °C to 2 °C by 2030 and of 3 °C to 6 °C by 2070 above 1990 levels under the "challenging" scenario. • Decrease in precipitation across southern Australia during winter and over southern and eastern Australia during spring over the coming decades. • Runoff affected by declining rainfall in winter, along with higher temperature. Rainfall and runoff changes will affect aquifer recharge. • Adverse impacts on the quality of surface resources and groundwater; possible contamination of water supply. • Increase in extreme rainfall events in many areas, increasing the likelihood and severity of floods. • Changes in the frequency of drought ranging from -20% to +80% by 2070 relative to present conditions, depending on the location. The largest increases in drought frequency are projected to take place in southwest Western Australia. • Increases in drought, salinity and other water quality issues and changes in water availability are threats to ecosystems. • The El Niño-Southern Oscillation (ENSO) phenomenon plays an important role in patterns of rainfall over parts of Australia, however potential changes in ENSO are not well understood.				
Primary concerns	Water quantity	Water quality	Water supply and sanitation	Extreme weather events	Ecosystems
	✓	✓ (e.g. salinity, exposure of acid sulphate soils in wetlands, blackwater events, increased toxic algal blooms)	✓ (increased development of WSS infrastructure – notably desalination plants and pipelines – in response to the Millennium Drought and threat of increased drought frequency)	✓ (drought, particularly in the south-west and floods, in some parts of northern Australia)	✓
Key vulnerabilities	• Australia is the driest of all inhabited continents. • Extreme variability of inter-annual rainfall in areas of high population, agricultural and environmental significance.				

Sources: Australian Government (2009), *Australia's Fifth National Communication on Climate Change*, Department of Climate Change, Canberra, *http://unfccc.int/national_reports/annex_i_natcom/submitted_natcom/items/4903.php*, accessed 15 June 2012; Australian Government (2012), *Water Policy and Climate Change in Australia*, National Water Commission, Canberra; Commonwealth Scientific and Industrial Research Organisation (CSIRO) and Australian Bureau of Meteorology (2012), *State of the Climate – 2012*, *www.csiro.au/Outcomes/Climate/Understanding/State-of-the-Climate-2012.aspx*, accessed 10 October 2012); PMSEIC Independent Working Group (2007), "Climate Change in Australia: Regional Impacts and Adaptation – Managing the Risk for Australia", *Report prepared for the Prime Minister's Science, Engineering and Innovation Council*, Canberra.

Key policy documents

Document	Reference to water?	Type	Year	Responsible institution
Water Act	Y	Legal act	2007	The Australian Government
Water Amendment Act	Y	Legal act	2008	The Australian Government
National Water Initiative	Y	National water policy framework	2004	The Council of Australian Governments
National Climate Change Adaptation Framework	Y	National adaptation strategy	2007	The Council of Australian Governments
National Coastal Risk Assessment Biodiversity Vulnerability Assessment	Y	National risk/vulnerability assessment	2009	Department of Climate Change and Energy Efficiency
Murray-Darling Basin Agreement (under the Water Act)	Y	Sub-national responses	2008	The Australian, ACT, NSW, QLD, SA, and VIC Governments
Murray-Darling Basin Plan (under the Water Act)	Y	Legal act	2012	Murray-Darling Basin Authority

Policy instruments

Areas \ Policy mix	Regulatory instruments	Economic instruments	Information and other instruments
Water quantity	• The water market rules: Prohibit actions that prevent or unreasonably delay irrigators from transforming their irrigation rights into separate statutory water access entitlements, allowing them to be traded outside the irrigation district. • The water charge rules: Promote efficient water pricing and sustainable use of water resources and water infrastructure across the Murray-Darling Basin. • Murray-Darling Basin Plan: Sets enforceable limits on the use of water that reflect an environmentally sustainable level of use. These limits will be responsive to changes in water availability.	• Water trading under the National Water Initiative (NWI):[1] Transfer of water access entitlements (permanent) and seasonal water allocations (temporary) between different entities. Water trading allows scarce water resources to be transferred to their most productive uses and allows access to water resources to be reallocated over time in response to: – changing commodity prices; – changing environmental conditions; – changes to the size of cities and towns; and – changes in the availability of water, *www.environment.gov.au/water/publications/action/nwi-wts-report.html.*	• The National Water Initiative (NWI) is Australia's enduring blueprint for water reform; through it, governments across Australia have agreed on actions to achieve a more cohesive national approach to the way Australia manages, plans for, measures, prices, and trades water, *www.environment.gov.au/water/australia/nwi/index.html.* This includes best practice water pricing, *www.environment.gov.au/water/publications/action/nwi-pricing-principles.html.* • The National Water Market System (NWMS) Project (ongoing): Aims to ensure water trading systems and processes better meet the Water Market Performance Characteristics set by Council of Australian Government and National Water Initiative requirements, *www.nationalwatermarket.gov.au/site-information/index.html.* • Reform of drought-related programmes: In October 2012, Australian primary industries ministers agreed to a new package of measures to better support farmers and their families to prepare for future challenges. The focus of the package is on helping farmers to better plan and prepare for drought and other challenges rather than waiting until they are in crisis to offer assistance, *www.daff.gov.au/agriculture-food/drought/drought-program-reform.*
Water quality			
Water supply and sanitation			
Extreme weather events		• The Government recently undertook public consultations on a proposal to require all general insurers offering home and contents insurance to include flood cover in those insurance policies whilst giving the insurers the option of allowing consumers to opt out of purchasing flood cover. The Government has deferred consideration of this proposal awaiting the final report of the Productivity Commission on *Barriers to Effective Climate Change Adaptation.*	

Policy instruments *(cont.)*

Areas \ Policy mix	Regulatory instruments	Economic instruments	Information and other instruments
Ecosystems	• The Murray-Darling Basin Plan: Must include an environmental watering plan (EWP) to co-ordinate environmental water use. One of the three primary objectives of the draft EWP is "to ensure that water-dependent ecosystems are resilient to climate change and other risks and threats", *www.mdba.gov.au/draft-basin-plan/supporting-documents/ewp.*	• Water entitlement buy backs: AUD 3.1 billion for purchasing water entitlements to help restore the health of vitally important rivers, wetlands, and floodplains, *www.environment.gov.au/water/policy-programs/entitlement-purchasing/index.html.*	• National Guidelines for the National Flood Risk Information Program: The Government will establish Guidelines for the reporting of flood risk information and establish a flood risk information portal to provide a single access point for flood mapping data, *www.em.gov.au/Publications/Program%20publications/Pages/National-Guidelines-for-the-National-Flood-Risk-Information-Program.aspx.* • Water for the Future programme: The Australian Government's effort to facilitate adaptation to climate change and as a response to increasing water scarcity (see below), *www.environment.gov.au/water/australia/index.html.* • The *National Water Quality Management Strategy* (NWQMS) provides information and tools to help communities manage their water resources to meet current and future needs,[2] *www.environment.gov.au/water/policy-programs/water-quality/index.html.* • Water Quality Improvement Plans (ongoing),[3] *www.environment.gov.au/water/policy-programs/nwqms/wqip/index.html.* • Australian Water Resources Information System (AWRIS): AWRIS is being developed by the Bureau of Meteorology to provide key water information to enhance understanding of Australia's water resources. • Water Efficiency and Labelling Standards (WELS) Scheme,[4] *www.waterrating.gov.au.*

1. Creating an environment in which individual water access entitlement holders are able to trade water quickly and easily will contribute to a more productive and efficient use of Australia's water over time. The Australian Government is also introducing water market charges and trading rules under the Water Act 2007, which will improve the water market by freeing up and setting rules for trade, and by ensuring appropriate price signals. Best practice water pricing is a key element of the National Water Initiative.
2. It provides policies, a process and a series of national guidelines for water quality management. As part of the NWQMS, the Water Quality Management Framework includes a step-by-step approach to planning, implementing and managing water quality, plus information about common environmental stressors.
3. The Australian Government is working in collaboration with States and Territories to develop Water Quality Improvement Plans (WQIP) to reduce pollution being released into freshwater ecosystems considered hotspots. It aims to reduce pollutants through the identification of environmental values of water, setting water quality and environmental flow objectives and implementation of catchment based management actions and monitoring.
4. It encourages water conservation by requiring that certain products, including clothes-washing machines, showers and lavatory equipment, be registered and labelled with their water efficiency when offered for supply.

Main research programmes

- The Australian Climate Change Science Program (ACCSP), delivered through the Australian Government Department of Climate Change and Energy Efficiency, the Commonwealth Scientific and Industrial Research Organisation (CSIRO), the Bureau of Meteorology and Australian universities, nationally co-ordinated under *A Plan for Implementing Climate Change Science in Australia.* The ACCSP improves understanding of the drivers of rainfall and how they are changing informing planning and adaptation decisions.
- The CSIRO research programme on climate change, including the Adaptation Flagship, *www.csiro.au/en/Outcomes/Climate/Understanding.aspx.*
- National Climate Change Adaptation Research Facility (NCCARF) Water Sector Adaptation Research Network, *www.nccarf.edu.au.*

Principal financing mechanisms and investment programmes

- The Australian Government has invested over AUD 300 million over the period 2007-12 to fund national adaptation actions, with those elements most relevant to water including:
  - AUD 50 million to establish and fund a National Climate Change Adaptation Research Facility (NCCARF) to build capacity in the research community on adaptation and to generate the information decision-makers need to adapt to the impacts of climate change, *www.nccarf.edu.au.*
  - AUD 44 million for a CSIRO Climate Change Adaptation Flagship, *www.csiro.au/en/Organisation-Structure/Flagships/Climate-Adaptation-Flagship/ClimateAdaptationFlagshipOverview.aspx.*
  - AUD 130 million to assist farmers to tackle the threat of climate change as part of the "Australia's Farming Future" program, *www.daff.gov.au/climatechange/australias-farming-future.*
  - Grants to local governments to develop strategies for managing risks from climate change impacts, *www.climatechange.gov.au/government/initiatives/lapp.aspx;www.climatechange.gov.au/government/initiatives/coastal-adaptation-decision-pathways.aspx.*
- Water for the Future programme: The Australian Government is investing AUD 14 billion over 12 years to facilitate adaptation to climate change and as a response to increasing water scarcity. The program aims to improve water management and deliver a range of water policy reforms. Measures under the program include:
  - Sustainable Rural Water Use and Infrastructure Programme: As part of the national investments in the Water for the Future initiative, the Australian Government has committed AUD 5.8 billion to increase water use efficiency in rural Australia. Under the programme, the Government has agreed to provide approximately AUD 3.2 billion for State Priority Projects (which includes Commonwealth led projects) in South Australia, New South Wales, Victoria, Queensland and the Australian Capital Territory (ACT), subject to a due diligence assessment of the social, economic, environmental, financial and technical aspects of the projects, *www.environment.gov.au/water/programs/srwui/index.html.*
  - Restoring the Balance in the Murray-Darling Basin Program: The Government has committed AUD 3.1 billion to facilitate water buy backs to protect and restore the environmental health of the Murray-Darling Basin. The programme allows irrigators to voluntarily sell their water entitlements to the Australian Government, *www.environment.gov.au/water/policy-programs/entitlement-purchasing/index.html.*
  - Urban water programmes to secure and diversify urban water supplies: Through the National Water Security Plan for Cities and Towns programme, the Australian Government has committed over AUD 250 million to fund practical projects that save water and reduce water losses in cities and towns nationally with populations of less than 50 000. In addition, over AUD 680 million has been committed to investments in desalination plants, water recycling schemes, stormwater harvesting and reuse projects. Funding is also provided for research into improving the technologies and use of desalination and water recycling, *www.environment.gov.au/water/policy-programs.*
  - Improving Water Information Program: The Bureau of Meteorology received funding of AUD 450 million over ten years to revolutionise the way water information is measured, accounted for, reported, forecasted and analysed.

Highlights and innovative initiatives

- **National Water Initiative:** In 2004, The Council of Australian Governments (COAG) agreed on a policy blueprint, the National Water Initiative, to improve the way Australia manages its water resources. It is designed to increase the efficiency of Australia's water use, leading to greater certainty of investment and productivity, for rural and urban use that optimises economic, social and environmental outcomes, *www.nwc.gov.au/reform/nwi.*
- **Water trading in the southern Murray-Darling Basin:** During severe drought conditions, water trading was vital in securing critical supply needs. The increases in flows due to trading were beneficial to river systems and provided important benefits to regions and local communities, *www.nwc.gov.au/publications/topic/rural/impacts-of-water-trading-in-the-southern-murray-darling-basin.*
- **The Commonwealth Environmental Water Holder:** The Commonwealth Environmental Water Holder is a statutory position under the Water Act that manages water recovered by the Australian Government to protect and restore environmental assets. This water must be managed in accordance with the EWP, *www.environment.gov.au/ewater/about/index.html.*

Austria

Climate change impacts on water systems

Observed changes and trends	• Increase in mean annual temperature by 1 °C to 2 °C within the last 50 years. • Increase in fall and winter precipitation in the northern part of Austria on a seasonal basis, while the southern part has seen a decrease in precipitation and run-off. • Decrease in the share of snowfall in total precipitation in high altitudes and glacier inventories show losses in area and volume.
Projected impacts	• Shifts in the seasonal precipitation distribution and enhanced water run-off in winter due to a higher fraction of rainfall (rather than snow). These shifts will lead to a more uniformly distributed annual cycle of water run-off. In general, this favors the production of hydropower, as supply will be more in phase with electricity demand. However, possible increases in sediment loading would impair the operation of hydropower infrastructure. • Shifts in the seasonal cycle of the precipitation also in the Alpine region, with an increase in precipitation during winter and a decrease in summer. However, a decrease in the annual amount of rainfall is projected only in the flat basins of southern and eastern Austria. • Reduction in the duration of snow cover due to changing precipitation regimes, altering the timing and amplitude of runoff from snow, increasing evaporation, and decreasing soil moisture and groundwater recharge. • Reduction of average precipitation and enhanced evapotranspiration due to higher than average temperatures can lead to deficiencies in drinking water supply, especially in communities depending on small and shallow springs in the southeast of Austria. • Offsetting the impact of climate change on snow coverage and duration via artificial snow making will place additional demand on water resources and energy. • Changes in the intensity and frequency of precipitation, temperature increase, glacier retreat and degradation of mountain permafrost can affect the frequency of natural hazards such as landslides, mudslides and avalanches. • Increases in extreme precipitation particularly in winter in the alpine region. • South-eastern Austria provinces of Burgenland, Styria and Carinthia might be affected by longer drought periods. • Changes in temperature and precipitation may lead to vegetation shifts and in some instances to extinction of species.

Primary concerns	Water quantity	Water quality	Water supply and sanitation	Extreme weather events	Ecosystems
	✓ (decrease in groundwater recharge and soil moisture)		✓ (drinking water supply)	✓ (heat waves (all cities), extreme precipitation, and longer drought periods)	✓ (mountainous regions, Lake Neusiedl)

Key vulnerabilities	• Austria is expected to be very vulnerable to climate change given that ecosystems in mountainous regions are highly sensitive. 70% of Austria's surface area is situated higher than 500 meters above sea level and 40% higher than 1 000 meters. • Longer droughts in the south-eastern provinces of Burgenland, Styria and Carinthia. • Lake Neusiedl, a famous bird breeding region, has been identified as a highly sensitive hydrological system, as the water balance of this lake is dominated by precipitation on and evaporation from the lake.

Sources: Austrian Federal Government (2009), *Fifth National Communication of the Austrian Federal Government*, Ministry of Agriculture, Forestry, Environment and Water Management, *http://unfccc.int/national_reports/annex_i_natcom/submitted_natcom/items/4903.php* (accessed 16 June 2012).

Key policy documents

Document	Reference to water?	Type	Year	Responsible institution
National Adaptation Strategy (available in German)[1]	Y	National adaptation strategy	2012	Ministry of Agriculture, Forestry, Environment and Water Management
Action Plan: Recommendations for implementation		National adaptation plan	2012	Ministry of Agriculture, Forestry, Environment and Water Management
Strategies for adaptation of Austrian Water Resources Management to Climate Change[2]	Y	National water strategy	2008	Task group[3] on behalf of the Federal Ministry of Agriculture, Forestry, Environment and Water Management in Co-operation with Austrian Water and Waste Association (ÖWAV)
Danube Strategy	Y	Transboundary responses[4]	Ongoing	

1. The Austrian National adaptation strategy consists of two parts: a general adaptation framework ("context") and a catalogue of adaptation options for 14 sectors/themes based on a qualitative vulnerability assessment (Action-Programme, NAP). Water is one of these 14 sectors (*www.lebensministerium.at/umwelt/klimaschutz/klimapolitik_ national/anpassungsstrategie/strategie-aussendung.html*).
2. See *www.lebensministerium.at/publikationen/wasser/hydrographischer_dienst/auswirkungen_des_klimawandels_auf_die_oesterreichische_wasser wirtschaft.html* (in German).
3. The relevance of effects of climate change to water and the need of adaptation has been recognised rather early in Austria, while the mainstream was still focusing more or less exclusively on mitigation. In 2007, a task group was initiated with the intention to identify possible impacts of climate change on Austrian water resources management and the need for actions in terms of developing strategies for adaptation of water resources management to expected long-term changes.
4. Austria has been an active proponent of the Danube Strategy, the macro-regional strategy for transnational territorial co-operation in the Danube region. Adaptation is a cross-cutting topic of relevance within a number of thematic priorities of the Danube Strategy. Implementation projects are currently in preparation.

Policy instruments

Areas \ Policy mix	Regulatory instruments	Economic instruments	Information and other instruments
Water quantity			• *www.klimawandelanpassung.at* provides information about climate change in Austria, adaptation to climate change, and activities towards a National Adaptation Strategy, *www.klimawandelanpassung.at/en.*
Water quality			
Water supply and sanitation			
Extreme weather events			
Ecosystems			

Main research programmes

- FLOODRISK I and II: Future-oriented implementation strategies for the integrated flood management (*www.umweltbundesamt.at/fileadmin/site/umweltthemen/klima/FloodRisk/FRII_english-abstracts.pdf*).
  - Research programmes providing knowledge base on climate change impacts as well as decision support to policy and administration in different sectors (but no specific focus on water): ACRP (Austrian Climate Research Programme) of the Klima- und Energiefonds (*www.klimafonds.gv.at/foerderungen/aktuelle-foerderungen/2011/austrian-climate-research-program*).
  - StartClim, *www.austroclim.at/index.php?id=40* (in German).
  - Mainstreaming of climate change adaptation in a range of other national research programmes, including provision, *www.provision-research.a* (available both in English and German); Global Change ÖAW, *www.oeaw.ac.at/deutsch/forschung/programme/programme.html* (available both in English and German), etc.
  - Assignment of specific policy support studies in sector policy fields (e.g. water).
  - Austrian participation in climate change projects under various EU funding schemes[1] (FP6/7, ESPON, European Territorial Co-operation, etc.).
- A scientific group of leading Austrian Research Institutes of Meteorology and Hydrology was commissioned to elaborate more in detail trends in historical data based on existing data and results of research projects, and to derive projections of possible climate change impacts. Based on these projections, strategies for adaptations in water resources management have been elaborated, available at: *www.lebensministerium.at/publikationen/wasser/wasserwirtschaft_wasserpolitik/Anpassungsstrategien-an-den-Klimawandel--Kurzfassung.html%20–%20Executive%20summary* (in German).
- On behalf of the Federal Ministry of Agriculture, Forestry, Environment and Water Management in Co-operation with Austrian Water and Waste Association (ÖWAV), the task group has published a report "Strategies for adaptation of Austrian Water Resources Management to Climate Change" in 2008, *www.lebensministerium.at/publikationen/wasser/hydrographischer_dienst/auswirkungen_des_klimawandels_auf_die_oesterreichische_wasserwirtschaft.html* (in German).

1. There are several projects under EU funding schemes, such as: "Adaptation to climate change in the Alpine Space" and "Water Management Strategies against water scarcity in the Alps" under ETC Alpine Space 2007-13 funding programme; "Adaptive management of climate-induced changes of habitat diversity in protected areas", "integrated approach to flood risk management" under "ETC Central Europe" funding programme; "Climate change and impacts on water supply" under the ETC South-East Europe funding programme.

Principal financing mechanisms and investment programmes

Belgium

Climate change impacts on water systems

Observed changes and trends	• Significant temperature rise in both summer and in winter by 2050. Rise in average summer temperatures ranging from 1.5 °C to 7 °C by the end of the century, as compared to the end of the 20th century. • Increased costs and damage linked to flooding in recent decades mainly due to the reduction of ground surface permeability, as well as construction of buildings in flood prone areas. • Increased frequency of surface water scarcity during periods of prolonged drought.
Projected impacts	• Increase in winter precipitation of 3% to 30% by the end of the century. Precipitation changes in summer varying between the current levels and a decrease of up to 50%. • An increase in winter precipitation will contribute to increased water recharge. • Reduced summer flows may negatively impact surface water quality and availability. • Increase of the number of heavy rainfalls episodes.

Primary concerns	Water quantity	Water quality	Water supply and sanitation	Extreme weather events	Ecosystems
	✓ (decrease in summer precipitation[1])	✓	✓	✓ (floods and droughts)	✓

Key vulnerabilities	• Metropolitan and coastal areas (such as the 65 kilometre-long coast and the Scheldt estuary) are vulnerable to flood risks. • Concerning water scarcity, in contrast with flood risks, less is known about the geographic distribution of the risk of water scarcity in periods of prolonged droughts. • The process of salinisation in the western province (West-Vlaanderen) due to overexploitation and the sea level rise is also a major concern in terms of water scarcity.

1. The expected decrease in summer precipitation can lead to a lack of water availability and problems for different sectors (e.g. agriculture, waterway transport).

Sources: National Climate Commission (2009), *Belgium's Fifth National Communication under the UNFCCC*, Federal Public Service Health, Food Chain Safety and Environment, Belgium, *http://unfccc.int/national_reports/annex_i_natcom/submitted_natcom/items/4903.php* (accessed 20 June 2012); ECORES-TEC (2011), *L'adaptation au changement climatique en Région wallonne*, *http://orbi.ulg.ac.be/handle/2268/113405* (accessed 21 October 2012).

Key policy documents

Document	Reference to water?	Type	Year	Responsible institution
Belgium:				
Belgian National Climate Change Adaptation Strategy	Y	National adaptation strategy	2010	National Climate Commission
Belgian National Climate Change Adaptation Plan	Y	National adaptation plan	Under development	National Climate Commission
Flemish Adaptation Plan	Y	Sub-national responses	Under development	LNE
Walloon Adaptation Plan	Y			AWAC
Brussels Adaptation Study	Y			IBGE- BIM

Key policy documents (cont.)

Document	Reference to water?	Type	Year	Responsible institution
Flanders: Sigma Plan[1]	Y	Transboundary responses	2005	Flemish Ministry for Mobility and Public Works (Agency "Waterwegen en Zeekanaal N.V. (W&Z)
Master Plan Coastal Safety (IMCORE)	Y	Sub-national and transboundary responses	2011	Agency "Maritieme Dienstverlening en Kust (MDL)"
First generation River basin management plans (RBPMs)	Y	Sub-national responses	2009	(Flemish) Coordination Commission on integrated Water Policy (CIW)
Water policy document including significant water management issues	Y	Sub-national and transboundary responses	Adoption foreseen in 2013	
Second generation RBMP	Y		Under development[2]	
Wallonia: First generation River basin management plans (RBPMs)	Y	Sub-national and transboundary responses	Under development	Directorate-General for Agriculture, Natural Resources and Environment
Brussels: First generation River basin management plans	Y	Sub-national and transboundary responses	Under development	IBGE-BIM

1. For more information, see *www.sigmaplan.be/en* (in English).
2. Adoption expected in 2015.

Policy instruments

Areas / Policy mix	Regulatory instruments	Economic instruments	Information and other instruments
Water quantity	**Flanders:** • "Water toets": A requirement that building plans are scrutinised for their potential impacts on water. The procedure applies in particular to flood prone areas, zones important for infiltration, or near drinking water catchment areas. • An additional measure in the first River Basin Management Plans defined a surface water abstraction permission that allows limiting or suspending the abstraction in periods of prolonged drought and low flows. • A permitting system for the abstraction of groundwater (> 500 m^3/year) that takes the quantitative status of the groundwater system into account.	**Flanders:** Groundwater abstraction charge: The price is differentiated by the aquifer and a regional factor.[1] • Surface water abstraction from navigable waterways: Abstraction charge for abstractions > 500 m^3/year.	**Belgium:** • Information campaign to promote water savings (launched in 2000). • Obligation for notaries and real estate agents to provide information about the risks of flooding of houses and building land. **Flanders:** • A flood warning and decision support system to forecast high flow events for the un-navigable watercourses 48 hours in advance, *www.overstromingsvoorspeller.be/default.aspx?KL=en* (in English). • Water metering obligation for ground water abstraction (except for domestic use < 500 m^3/year). **Wallonia:** Hydrological instruments in Wallonia: • For navigable watercourses: *http://voies-hydrauliques.wallonie.be/opencms/opencms/fr/hydro/Actuelle/crue/index.html* (in French). • For non-navigable watercourses: *http://aqualim.environnement.wallonie.be* (in French). • Real-time monitoring of watercourses, hydrology studies and co-ordination of flood alert.
Water quality	**Flanders:** Discharge permits. **Brussels:** • Brussels Water Management Plan: Adopted in 2012, contains measures about drinking water quality draining, wastewater treatment and watercourses protection, *http://documentation. bruxellesenvironnement.be/documents/Plan_Eau_PGE_7_Programme_de_Mesures_2012_FR.PDF?langtype=2060* (in French).		

Policy instruments (*cont.*)

Areas \ Policy mix	Regulatory instruments	Economic instruments	Information and other instruments
Water quality *(cont.)*	• Brussels Adaptation study (2012): Provides measures aiming *inter alia* to optimise the surface water management in order to maintain it in a good quantitative and ecological condition.		
Water supply and sanitation	**Flanders:** • An additional measure in the first River Basin Management Plans to stimulate rational water use: Allows adjustments to the tariff structure for drinking water to distinguish basic uses and discretionary (or luxury) uses. **Brussels:** • Water Management Plan (see above). • Brussels Adaptation study: Provides measures aiming to reinforce the control of groundwater as well as the supply and demand of water in a climate change perspective.	• Charges for use, taxes, purification, etc.	
Extreme weather events	**Wallonia:** Flood prevention plan in Wallonia (PLUIES plan): Approved in 2003, aims to improve knowledge of the risk of flooding, reduce and decelerate run-off, improve river management and reduce vulnerability of flood-prone areas, *http://environnement.wallonie.be* (in French). **Brussels:** • Flood prevention plan in Brussels (*Plan Pluie*, approved in 2008): Promotes preventive measures to ensure the built environment is better adapted to increased precipitation. Measures are being implemented to recover rain water (water tanks are compulsory for new housing and promoted through regional grants for existing housing) and increase evapotranspiration (e.g. limit on built-up areas, green roofs), *http://documentation.bruxellesenvironnement.be/documents/Plan_pluie_2008-2011_RIE_FR.PDF* (in French). • Adaptation to drought risks and heat waves, and improvement of urban micro-climates (characterised by the "heat island" effect), through the measures of the Water Management Plan.	**Belgium:** Insurance for floods and other natural hazards: Recent reforms[2] at the federal level introduced coverage for natural disasters in household fire insurance policies. In a departure from previous arrangements, state subsidies to support coverage for natural disaster will not be provided.[3]	
Ecosystems			

1. The latter factor will annually rise from 2010 to 2017 to take into account the pressure on groundwater and to stimulate rational water use.
2. Laws of 21 May 2003 and 17 September 2005. While not primarily targeted at adaptation, the new laws may discourage new residential construction in areas where the risk of flooding is higher, especially if this results in higher insurance premiums. There is also a price setting board to limit the premiums for existing constructions in high-risk areas, by sharing the cost among all insured parties. However, new construction in high risk zones would be excluded from the premium limitation mechanism, effectively making such constructions likely uninsurable.
3. With the exception of situations where the total cost of cover against natural disasters exceeds a threshold linked to the turnover insurance companies.

Main research programmes

- Research Program "Science for a Sustainable Development": Belgian Science Policy has undertaken research on climate change and inland floods, *www.belspo.be/belspo/ssd/index_en.stm* (in English).
- Research report "Climatic Vigilance" (2009) by the Royal Institute of Meteorology: Described observed climatic trends from 1830 to present day, *www.meteo.be/meteo/view/fr/66940-Artikels.html?view=3236558* (in French).
- Support of research and development projects by the agency for innovation by Science and Technology (IWT): Supporting the research project CcASPAR on spatial planning, *www.iwt.be/english/funding* (in English).
- European Research Project AMICE "Adaptation of the Meuse river to the Impacts of Climate Evolution", *www.amice-project.eu/en* (in English).

Principal financing mechanisms and/or investment programmes

Highlights and innovative initiatives

- **Green spatial planning policy** to reduce inundations by reducing paved surface, *www.vlaanderen.be/ruimtelijk/br2012/groenboek_beleidsplanruimte.pdf* (in Dutch).
- **Various national and international projects on climate change**, *www.lne.be/themas/klimaatverandering/adaptatie/studies-en-onderzoek/studies-en-onderzoeken* (in Dutch).
- In Flanders, the Water Framework Directive and the Flood Directive are transposed via a single legislative instrument – the Decree on Integrated Water Management. The Flood Risk Management Plan will be integrated in the River Basin Management Plan in 2015.

Canada

Climate change impacts on water systems

Observed changes and trends	• Climate change impacts on water resources in Canada vary regionally including shortages (droughts), excesses (floods) and associated water quality issues, depending on the season. • Since 1948, snowfall has increased in the North, and decreased in South-Western Canada. • Glacial retreat is widespread since the late 1800s in Western Canada and since the 1920s in the Arctic. • Decline in summer and fall runoff in the Prairies provinces, leading to seasonally lower lake and river levels. • Northward or upslope shifts in terrestrial ecosystems, shifts towards warmer thermal regimes in freshwater ecosystems. • On average, Great Lakes water levels are currently about 1 metre below the long-term average.
Projected impacts	• Increase in annual total precipitation across the country during the current century. However, throughout most of southern Canada, precipitation increases are projected to be low (0% to 10% by 2050s) during the summer and fall months. In some regions, especially in the south-central Prairies and southwestern British Columbia, precipitation is expected to decline in the summer. • Increase in the frequency of water shortages in Ontario, British Columbia, and in the Prairie provinces. • Increase in surface water temperatures, decrease in the duration of ice cover, and lower water levels due to warmer conditions. These changes lead to higher pollutant concentrations. Increased flooding will also contribute to water quality degradation. • Water supply issues to become a greater concern in the Great Lake Basin due to lower lake levels, which may reduce hydroelectricity production, harm tourism and recreation, and force vessels to decrease cargo capacity. • Increase in the frequency, intensity or duration of extreme weather/climate events in most regions. • Increase in the risk of flooding in many regions in winter due to less ice cover, more precipitation events, and more frequent winter thaw. An increase in extreme daily precipitation is expected. • More frequent and sustained drought in British Columbia. • Adverse effects on ecosystems in Québec, Ontario, and in the Prairie provinces. In particular, changes in winter snowfall will have an adverse impact on Prairie wetlands.

Primary concerns	Water quantity	Water quality	Water supply and sanitation	Extreme weather events	Ecosystems
	✓ (Seasonal and regional basis)	✓ (Seasonal and regional basis)		✓	

Key vulnerabilities	• Canada's Arctic is particularly vulnerable to climate change. Infrastructure in the Arctic regions has already been exposed to significant impacts that require urgent changes to engineering practices, climatic design information, codes and standards from melting permafrost, warming temperatures, changing snow loads and other extreme precipitation. • Water-stressed areas will expand due to decreased runoff in many areas, while reduced water quality and quantity will be experienced on a seasonal basis in every region of Canada.

Sources: Environment Canada (2012), *Climate Information to Inform New Codes and Standards*, *www.ec.gc.ca/sc-cs/default.asp?lang=En&n=20CD1ADB-1* (accessed 2 October 2012); Environment Canada (2004), *Threats to Water Availability in Canada*, National Water Research Institute, Burlington, Ontario, NWRI Scientific Assessment Report Series No. 3 and ACSD Science Assessment Series No. 1, *www.ec.gc.ca/inre-nwri/default.asp?lang=En&n=0CD66675-1&offset=1&toc=show* (accessed 20 November 2012); Government of Canada (2010), *Fifth National Communication on Climate Change*, *http://unfccc.int/national_reports/annex_i_natcom/submitted_natcom/items/4903.php* (accessed 20 June 2012); Government of Canada (2007), *From Impacts to Adaptation: Canada in a Changing Climate*, D.S. Lemmen et al. (eds.), Natural Resources Canada, Environment Canada, Ottawa, Ontario; Government of Canada (2004), *Climate Change Impacts and Adaptation: A Canadian Perspective*, D.S. Lemmen and F.J. Warren (eds.), Natural Resources Canada, Ottawa, Ontario.

Key policy documents

Document	Reference to water?	Type	Year	Responsible institution
Federal Adaptation Policy Framework	N	Federal adaptation framework	2011	Government of Canada
Canada Country Study: Climate Impacts and Adaptation	Y	National impact assessment	2000	Environment Canada
From Impacts to Adaptation: Canada in a Changing Climate	Y	National risk assessment	2008	Natural Resources Canada
Strategic Directions for Water: Three Year Action Plan[1]	Y	Water strategy document	2010	Canadian Council of Ministers of the Environment
Climate Ready: Ontario's Adaptation Strategy and Action Plan	Y	Sub-national responses	2011	Government of Ontario
The Water Opportunities Act	Y	Sub-national responses	2010	Government of Ontario
Preparing for Climate Change: British Columbia's Adaptation Strategy	Y	Sub-national responses	2010	Government of British Columbia
Living Water Smart	Y	Sub-national responses		Government of British Columbia
Water for Life: Nova Scotia's Water Resource	Y	Sub-national responses	2010	Government of Nova Scotia
Manitoba Water Strategy	Y	Sub-national responses	2003	Government of Manitoba
Northwest Territories Water Stewardship Strategy	Y	Sub-national responses	2010	Government of the Northwest Territories
Water for Life	Y	Sub-national responses	2008	Government of Alberta
Water for Life: Action Plan	Y	Sub-national responses	2009	Government of Alberta
Climate Change Action Plan	Y	Sub-national responses	2008	Government of Québec
Quebec Water Policy	Y	Sub-national responses	2002	Government of Québec
Saskatchewan Energy and Climate Change Plan	Y	Sub-national responses	2007	Government of Saskatchewan
Pan-Territorial Adaptation Strategy: Moving Forward on Adaptation in Canada's North	Y	Sub-national responses	2011	The Governments of Nunavut, the Northwest Territories and Yukon
Climate Change Adaptation Strategy for Atlantic Canada	Y	Sub-national responses	2008	Governments of New Brunswick, Prince Edward Island, Nova Scotia, and Newfoundland and Labrador
Climate Change Action Plan	Y	Sub-national responses	2007	Government of New Brunswick
Prince Edward Island and Climate Change: A Strategy for Reducing the Impacts of Global Warming	Y	Sub-national responses	2008	Government of Prince Edward Island
Climate Change Action Plan	Y	Sub-national responses	2005	Government of Newfoundland and Labrador

1. One of the goals is to reduce climate change impacts on water through adaptive strategies.

Policy instruments[1]

<table>
<tr><th>Areas \ Policy mix</th><th>Regulatory instruments[2]</th><th>Economic instruments</th><th>Information and other instruments</th></tr>
<tr><td>Water quantity</td><td>Ontario:
The Low Water Response Programme: Aims to reduce the impacts of drought on water supplies. It was developed in response to the increased occurrence of low water levels due to changing weather patterns, such as a severe drought in 1998-99.</td><td></td><td rowspan="5">• Innovative risk management tools for adaptation: Guidelines and tools to address risks from climate change, e.g. "Adapting to Climate Change: A Risk Based Guide for Local Governments", "Infrastructure Climate Risk Protocol", www.nrcan.gc.ca/earth-sciences/climate-change/community-adaptation/295.
• Adaptation planning tools: Guide users from risk assessment to the identification and implementation of adaptation responses, e.g. "Changing Climate, Changing Communities: Guide for Municipal Climate Adaptation" and "Canadian Communities' Guidebook for Adaptation to Climate Change", www.nrcan.gc.ca/earth-sciences/climate-change/community-adaptation/295.
• Water and Climate Compendium: A website providing access to information, knowledge products, and training on water resource adaptation to climate change, http://waterandclimate.ca/WP.
• Early warning system: Environment Canada's Meteorological Service has the unique mandate and capacity to warn Canadians of impending high impact weather events that can affect their lives and property.
• A Canada-wide Framework for Water Quality Monitoring (2006): A guide for jurisdictions in the planning and implementation of water qualitymonitoring programs, www.ccme.ca/assets/pdf/wqm_framework_1.0_e_web.pdf.</td></tr>
<tr><td>Water quality</td><td></td><td></td></tr>
<tr><td>Water supply and sanitation</td><td>Ontario:
The Water Opportunities and Water Conservation Act (2010): Enables the province to develop regulations that will require new and innovative ways to reduce demands on existing water resources and also address impacts from a changing climate. The Act requires municipalities to create a Municipal Water Sustainability Plan that includes conservation plans, a risk assessment and risk management plan which address the challenges of climate change and impacts on water resources.</td><td></td></tr>
<tr><td>Extreme weather events</td><td>Ontario:
• The Building Code is an important policy tool in responding to the direct and indirect effects of climate change. Work is underway by the Ministry of Municipal Affairs and Housing (MMAH) on the development of the next edition of the Building Code to make new buildings in Ontario resilient to climate change impacts and to enhance their ability to conserve water and energy.</td><td></td></tr>
<tr><td>Ecosystems</td><td></td><td></td></tr>
</table>

1. Water management issues are primarily the responsibility of provincial and territorial governments in Canada. Major provincial and territorial initiatives relating to climate change adaptation and water are listed in Section 2. For more information on principal measures to address climate change adaptation for water systems at sub-national level, please see provincial and territorial government websites for further information.
2. This section provides some examples of regulatory instruments from the Government of Ontario.

Main research programmes

- Environment Canada undertakes a broad range of climate research including process studies, data analysis, greenhouse gas monitoring and global and regional climate modeling, *www.ec.gc.ca/sc-cs/default.asp?lang=En&n=1F788646-1.*
- Environment Canada also conducts national scale research on climate change impacts and water availability, *www.ec.gc.ca/sc-cs/default.asp?lang=En&n=9AF9494E-1.* Research programmes focussed on climate change and water are undertaken by several departments.
- The Natural Sciences and Engineering Research Council of Canada announced in 2012 a program called "Climate Change and Atmospheric Research' (CCAR) to fund new climate change research in Canada and foster improved collaboration between government and university researchers, *www.nserc-crsng.gc.ca/Professors-Professeurs/Grants-Subs/CCAR-RCCA_eng.asp.*

Principal financing mechanisms and investment programmes

- The Government of Canada invested CAN 85.9 million over the period 2007-11 to help Canadians increase their capacity to adapt to a changing climate. These initiatives enhanced the scientific knowledge and tools needed to take further action against climate change and reduce the risks to Canadians, *www.ec.gc.ca/default.asp?lang=En&xml=91E1F38E-C53C-404B-9512-22EA69C08787*.
- In November 2011, the Government announced CAN 148.8 million over five years for projects to improve the understanding of climate change impacts focusing on science to inform adaptation decision making, health, the North and Aboriginal communities, and economic competitiveness, *www.climatechange.gc.ca/default.asp?lang=En&n=2B2A953E-1* (note that not all of these programs focus on water issues).
- Funding for new climate change research in Canada will be provided by the Natural Sciences and Engineering Research Council of Canada under the new program "Climate Change and Atmospheric Research" (CCAR), *www.nserc-crsng.gc.ca/Professors-Professeurs/Grants-Subs/CCAR-RCCA_eng.asp*.

Highlights and innovative initiatives

- The **Regional Adaptation Collaboratives (RACs) Climate Change Program** was a three year, CAN 30 million, cost-shared federal program to help Canadians reduce the risks and maximise the opportunities posed by climate change. This Program helped communities prepare for and adapt to local impacts posed by our changing climate, such as decreasing fresh water supplies, increasing droughts, floods, and coastal erosion.
  - The British Columbia RAC, Preparing for Climate Change: Securing B.C.'s Water Future, focused on enhancing resilience to a changing climate and the anticipated, related impacts on water and aquatic ecosystems, *www.nrcan.gc.ca/earth-sciences/climate-change/community-adaptation/regional-collaborative/636*.
  - The Prairies RAC focused on a series of projects to advance decision making in the areas of water supply and demand, drought and flood planning, and, forest and grassland ecosystems, *www.nrcan.gc.ca/earth-sciences/climate-change/community-adaptation/regional-collaborative/175*.
  - To address targeted climate change vulnerabilities, the Ontario RAC conducted a series of projects to advance decision making in the areas of extreme weather risk management, water management and community development planning, *www.nrcan.gc.ca/earth-sciences/climate-change/community-adaptation/regional-collaborative/189*.
  - Water management was a focus of the Quebec RAC, which aimed to initiate adaptation on targeted issues by providing specialised information and tools to government water managers and watershed groups, *www.nrcan.gc.ca/earth-sciences/climate-change/community-adaptation/regional-collaborative/768#water*.
  - The Atlantic Climate Adaptation Solutions Project worked collaboratively with a broad range of government and non-government partners to enhance Atlantic Canada's resilience to a changing climate. A series of projects were undertaken to address vulnerabilities in this region, *www.nrcan.gc.ca/earth-sciences/climate-change/community-adaptation/regional-collaborative/204#enhance*.
- The **Northern Voices, Northern Waters** – Northwest Territory Water Stewardship Strategy: Released in May 2010, the Strategy aims to ensure the maintenance of the quantity and quality of water resources in the Northwest Territory, secure water supply, and the resilience of freshwater ecosystems. An action plan for the period of 2011-15 has been developed, *www.enr.gov.nt.ca/_live/pages/wpPages/water.aspx*.

Chile

Climate change impacts on water systems

Observed changes and trends	• Rise in temperature in the Central Valley and particularly the Andes Mountains (where most of Chile's water resources are stored) for the period 1979-2006. For the same period, temperatures in the ocean and on the coast have tended to drop. • Glaciers are in retreat.				
Projected impacts	• Increase in mean temperature for continental Chile by 2 °C to 4 °C, with greater increase in Andean regions and smaller increases toward the South. Only in southern Chile are temperatures projected to rise by less than 1 °C. Greater warming in summer, exceeding 5 °C in some areas of the high Andes. • Change in annual precipitation by more than 30% in some areas of the country by 2040. Central Chile (where 70% of the total population lives) may see significant reduction in precipitation. Decrease in precipitation of around 20 to 25% between Antofagasta and Puerto Montt, but an increase from Chiloé Island to the south. • Reduction in the mountainous area capable of storing snow over successive years and shift in snow line towards higher altitudes. • Retreat of glaciers will have a significant impact on water supply, as glaciers act as strategic water reserves, not only supplying water to river basins in summer, but providing the single most important source of replenishment for rivers, lakes, and groundwater in arid regions during periods of drought. • Decreasing available water flow in all river basins that have been assessed. Reductions will be greater in the most northern and southern regions (the Limarí and Cautín Basins) while the remaining basins are projected to experience slight reductions in flow levels in the short-term and significant reductions in the mid-term. • Major increase in the number of months with a hydrologic deficit in practically all river basins, due to the projected changes in availability and seasonal distribution of the water flows. This will greatly affect the availability of water resources, with low-flows occurring more frequently. • Decreasing ecosystem services of surface and ground water will have a significant impact on the quality of water recourses. In general, the water system's capacity to dilute and regulate pollutants and liquid waste emissions will be reduced. In addition, increased frequency of extreme events will degrade surface water quality. • Increase in drought, especially in the northern and central regions. River overflows across the country due to *El Niño* Southern Oscillation (ENSO) events. • Negative impacts on ecosystems that depend on the quality and quantity of water resources. Water quality affects the natural conditions that allow for the development of aquatic ecosystems and biodiversity. In addition, ecosystems that depend on groundwater supply will be affected.				
Primary concerns	Water quantity	Water quality	Water supply and sanitation	Extreme weather events	Ecosystems
	✓ (decrease in average annual river flows and recharge of groundwater; change in seasonal patterns of snow melt)	✓ (degradation of surface water quality due to extreme events; groundwater salinisation in coastal zones; degradation of groundwater quality in northern areas; decrease in the regulation ability of water systems, such as dilution of pollutants)	✓ (increased water scarcity and increased demand from other sectors can affect availability of drinking water)	✓ (drought, especially in northern and central basins; river overflows across the country due to ENSO events)	✓ (impoverishment of aquatic life and alteration of biodiversity depending on groundwater availability)
Key vulnerabilities	• Northern and central Chile where water is already scarce and where considerable reductions in rainfall are expected. • Chile has the highest continental concentration of glaciers in the Southern Hemisphere. In 2007, the country's 1 835 glaciers composed a total area of 15 500 km^2. • Chile has more than 20 000 km^2 of ice reserves, 75 % of which is found in the northern and southern Patagonian Ice Fields.				

Sources: Economic Commission for Latin America and the Caribbean (2012), *La Economía del Cambio Climático en Chile*, *www.eclac.cl/publicaciones/xml/0/47220/La_economia_del_cambio_climatico_en_Chile_Completo.pdf* (accessed 8 August 2012); Ministerio de Energía (2011), *Selección y Aplicación de un Modelo Hidrológico para estimar los Impactos del Cambio Climático en la Generación de Energía del Sistema Interconectado Central*, *www.minenergia.cl/documentos/estudios/seleccion-y-aplicacion-de-un-modelo.html* (accessed 10 September 2012); Ministerio del Medio Ambiente (2011), *Segunda Comunicación Nacional de Chile*, *http://unfccc.int/national_reports/non-annex_i_natcom/items/2979.php* (accessed 22 June 2012); National Environmental Commission (CONAMA) (2010), *Análisis de Vulnerabilidad de Recursos Hídricos frente a escenarios de Cambio Climático para las cuencas Cautín, Aconcagua, Teno e Illapel*, National Environmental Commission, Department of Climate Change; National Environmental Commission (CONAMA) (2008), *National Climate Change Action Plan for 2008-12*, National Environmental Commission, Department of Climate Change, *www.mma.gob.cl/1304/articles-49744_Plan_02.pdf* (accessed 15 September 2012).

Key policy documents

Document	Reference to water?	Type	Year	Responsible institution
National Climate Change Strategy	Y	National climate change strategy	2006	National Environmental Commission (CONAMA)
National Climate Change Action Plan for 2008-12	Y	National climate change action plan	2008	CONAMA
Eight National Adaptation Plans for key sectors (including water resources)	Y	National sectoral adaptation plan	Under development (2012-17)	Climate Change Office, Ministry of Environment
National Adaptation Plan[1]		National adaptation plan	Under development	Climate Change Office, Ministry of Environment
Water Resources Adaptation Strategy[2]	Y	Water resources adaptation strategy	Under development	Climate Change Office, Ministry of Environment
Glacier protection and conservation policy	Y		2009	General Directorate of Water of the Ministry of Public Works

1. It will propose the lines of action to address adaptation to climate change.
2. It will propose specific measures for water resources.

Policy instruments

Areas \ Policy mix	Regulatory instruments	Economic instruments	Information and other instruments
Water quantity	• Water Code (DFL 1122): Modification (pending approval) that introduces changes to the Water Code, increasing fines and imprisonment for illegal water extraction. Establishes penalties for noncompliance with the process of executing transfers of water rights, according to the responsible bodies (Notaries and Water Rights Conservatives). Improvements in supervision of extractions and water market are also expected.		• National Plan for Climate Change Education and Awareness: incorporating the subject into curricula at all educational levels (Ministry of Education, planned). • Improved prediction and response for floods. • National irrigation strategy includes a plan considering the construction of 15 reservoirs in the long term, and 4 during the present administration in order to expand irrigation. • The General Directorate of Water is conducting studies to provide information about the feasibility of desalination plants.
Water quality	• Emission Norm for regulation of the pollutants associated with liquid waste discharges to marine and continental surface water (DS 90/2000, of the Ministry of the Secretary General of the Presidency). • Emission Norm of liquid waste to ground water (DS 46/2002, of the Ministry of the Secretary General of the Presidency).[1]		
Water supply and sanitation	• Emission Norm for regulation of the pollutants associated with industrial liquid waste discharges to main drainage systems (DS 609/98, of the Ministry of Public Works) provides a quality standard to reduce the costs of water treatment. • Water Resources National Strategy: proposes (among other measures) the creation of new sources of water. This includes a plan of reservoirs to be addressed in the next 10 years and groundwater infiltration projects, with a first pilot project in the Aconcagua Valley. Both measures aim to mitigate water scarcity due to climate change.		

Policy instruments *(cont.)*

Areas \ Policy mix	Regulatory instruments	Economic instruments	Information and other instruments
Extreme weather events			
Ecosystems	• Quality Environment Secondary Norm for the water protection of the Llanquihue Lake (DS 122/10, of the Ministry of the Secretary General of the Presidency). • Quality Environment Secondary Norm for the protection of continental surface water of the Serrano River Basin (DS 75/10, of the Ministry of the Secretary General of the Presidency).[2]	• Currently (2012) the Ministry of Environment analyses the feasibility of a "compensation bank" as an instrument of public policy to compensate losses in biodiversity through the SEIA (Environmental Impact Evaluation system) framework.	

1. These norms are the current instruments that provide assurance in terms of water quality. They could help to address the effects of climate change on water quality due to decreases in flows.
2. Other secondary norms are in process of approval. These norms are instruments for protection of the ecosystems related to water bodies and could reduce the negative effects of climate change.

Main research programmes

- Glaciology and Snow Unit: created in 2008 within the Ministry of Public Works' General Directorate of Water. The Unit is intended to establish and implement a national glaciology programme that will develop a glacier inventory, study and monitor glaciers in Chile. It will also define responses to climate change for glaciers and identify adaptation strategies for different climate scenarios.
- The National Irrigation Commission supports research on climate change adaptation and water, focusing on irrigation and associated impacts, developed by universities and research centres in Chile with national and international funding. The main research topics are: 1) decision making support to reduce vulnerability of irrigated agriculture to climate variability and long-term change; and 2) the development of an adaptation process to address climate change impacts and vulnerability in water management at the basin level.

Principal financing mechanisms and/ or investment programmes

- The Global Environment Facility (GEF) and its implementing agencies, along with bilateral development co-operation partners, have provided funding and technical support to develop and strengthen activities related to climate change. The Government of Chile has provided funding for managing climate change, enabling the creation of permanent working groups charged with addressing climate change from within their ministries and the allocation of budgets to implement their activities.

Czech Republic

Climate change impacts on water systems

Observed changes and trends	• Increase in average annual total precipitation by 2.9% from 1991 to 2008, compared to the period from 1961 to 1990. • Greater occurrence of extreme weather events, especially in the last 10 years.				
Projected impacts	• Decrease in average flow rates is expected in the medium term in many river basins ranging from 15-20% ("optimistic" scenario) to 25-40% ("pessimistic" scenario). This will lead to fundamental changes in the overall hydrological regime. • Higher winter temperatures are projected to shift annual outflows due to a reduction (in some places very substantial) in snowpack and an increase in territorial evaporation. • Increase in winter runoff and increase in the risk of spring floods can be expected in parallel with reduced water storage in the form of snowpack. • Increase in the likelihood of a substantial reduction in the storage function of reservoirs. Large reservoirs are less sensitive to changes than smaller reservoirs. • Intense precipitation events that occur during summer thunderstorms will present a greater risk of flash floods, even in cases where long-term total precipitation remains stable.				
Primary concerns	Water quantity	Water quality	Water supply and sanitation	Extreme weather events	Ecosystems
	✓ (decrease in river flow/ groundwater level	✓ (increase in the concentration of pollutants)[1]		✓ (flash floods, local drought)	
Key vulnerabilities					

1. Depending on the river flow decreases, threat of eutrophication.

Sources: Ministry of Environment (2009) *Fifth National Communication of the Czech Republic on the UNFCCC,http://unfccc.int/national_reports/annex_i_natcom/submitted_natcom/items/4903.php* (accessed 22 June 2012); Ministry of Environment (2007) *National Program to Abate the Climate Change Impacts in the Czech Republic, www.mzp.cz/en/national_programme* (accessed 22 June 2012).

Key policy documents

Document	Reference to water?	Type	Year of adoption	Responsible institution
Water Act (ACT No. 254/2001 Coll. [4109, Head IX])	Y	Legal act	2001/2010	Ministry of Environment, Ministry of agriculture
National Program to Abate Climate Change Impacts in the Czech Republic	Y	National climate change programme	2004	Ministry of Environment
		National adaptation strategy	Under development	

Policy instruments

Areas \ Policy mix	Regulatory instruments	Economic instruments	Information and other instruments
Water quantity	• Revision of operating regulations of water infrastructures: Provisions to increase the safety from overflow, to improve the effectiveness of their management under non-stationary conditions, and for decision making processes in dangerous or uncertain situations.	• Payments for water use: In the context of adaptation to climate change, it is necessary to maintain optimal water stream flows. One possible solution to limit water withdrawals is to adjust fees for the withdrawal.	• Plan for main river basins:[1] Supports the implementation of adaptation measures, in particular, increasing the retention of water in the landscape and improving the structure of the landscape. • CzechGlobe: Responds to public demand for information by providing expert knowledge of global change issues and on the development of practices that lead to mitigating or to adapting to its effects, *www.czechglobe.cz/en/home.html.* • River Basin Management Plans and Flood Risk Management Plans preparation: RBMPs (2009) partly include the measures against climate change, however this issue will be more reflected in next RBMPs (2015).
Water quality	• Wastewater treatment requirements: Including emission limits and the regular revision of permits.		
Water supply and sanitation	• Revision of rules[2] on water mains and sewers to reduce water losses and minimise pollution: Introduced the obligation to prepare and implement Plans of financing (for at least 10 years) for renewal of water mains and sewers to reduce losses in drinking water mains, reduce demands on consumption of drinking water and minimise pollution of water courses through discharge of waste waters.		
Extreme weather events	• Prevention of Floods programme of the Ministry of Agriculture:[3] Aims to improve protection of the most endangered areas against floods, *http://eagri.cz/public/web/en/mze/water/flood-prevention.* • Flood Prevention Programme II:[4] Supports flood prevention measures with retention, along water courses, increasing the safety of water works and zoning of inundation areas (for the period 2007-12).	• Environment Operational programme.	
Ecosystems	• Minimal residual flow and minimal groundwater level (under development): Maintaining minimum residual flow is determined by the Water Act. The new government regulation with more strict conditions on maintaining the minimum residual flows is expected to be adopted next year. It should preserve ecosystem functions linked to freshwater system.		

1. Government Resolution No. 562/2007.
2. The 2006 amendment to Act No. 274/2001 Coll. on water mains and sewers for public use. The latest amendment is under preparation, which aims to simplify administrative procedures and eliminate deficiencies identified from experience.
3. Government Resolution No. 382/2000.
4. Ministry of Agriculture Decree No. 560/2006 Coll.

Main research programmes

- Research project on climate change impacts on hydrology, water management, agriculture and forestry and proposals for adaptation options (2007-11): Provided specification and update of climate change scenarios for the Czech Republic up to 2050, specify expected impacts of climate change in hydrology, water management, agriculture and forestry sectors, propose relevant adaptation options and support meeting of the National Climate Change Programme in the Czech Republic. *www.isvav.cz/projectDetail.do?rowId=SP%2F1A6%2F108%2F07.*
- Research project to evaluate influence of climate changes on the hydrological balance and propose practical measures to mitigate their impacts (2007-11): Specified actual estimations of climate change impacts on hydrological balance, extreme hydrological events and water sources in the Czech Republic, *www.isvav.cz/projectDetail.do?rowId=SP%2F1A6%2F151%2F07.*
- Czech Hydrometeorological Institute (CHMI): Provides information on climate change observations and projections, in addition to actual weather conditions and alerts for extreme events, *www.chmi.cz/portal/dt?portal_lang=en&menu=JSPTabContainer/P1_0_Home.*
- T.G. Masaryk Water Research Institute, public research institution: Temporal and Spatial Variability of Hydrological Draught in Climate Change Conditions on the Territory of the Czech Republic (2008-2010), *www.isvav.cz/projectDetail.do?rowId=SP%2F1A6%2F125%2F08.*

Principal financing mechanisms and investment programmes

- Prevention of Floods programmes I and II: During the first phase of the programme (2002-06), a total of CZK 3.629 billion was spent on anti-flood protection measures focussed mainly on the improvement of protection against floods in endangered areas, increasing the possibilities of water retention, the development of dams allowing harmless water overflow into river flood plains and on enlarging the capacity of river-beds. The second phase of the project "Flood Prevention Programme II" and "Support for Renewal, Dredging and Reconstruction of Fishponds and Construction of Water Reservoirs" followed. The total costs are estimated to be around CZK 15 billion. The project also includes the implementation of land adaptations aimed at flood prevention, *http://eagri.cz/public/web/en/mze/water/flood-prevention.*
- River System Restoration Programme (Ministry of Environment Directive No. 5/2006): Aims to promote the renewal of the natural function of watercourses and to increase the water retention ability of the landscape. This is predominantly an investment programme.
- Environment Operational Programme: Based on Europeans funds (Cohesion Fund and European Regional Development Fund) for 2007-13, provides co-financing for projects concerned with improving the water management infrastructure, reducing flood risk and improving the water regime in the landscape.

Highlights and innovative initiatives

- **Programme of renewal of the natural functions of the landscape:** provides support for adaptation measures contributing to improving the natural function of water courses, improving the natural retention capacity of the landscape and the establishment and restoration of elements of ecosystems influencing the water regime. A subsidy of up to 100% of the total costs of the project is available. Tens of millions of Czech crowns are allocated annually. The program is divided into several sub-programs:
  - ensuring responsibilities of nature protection authorities;
  - implementation and preparation of plans and programs for protected species of plants and animals;
  - adaptation measures to mitigate the impacts of climate change on aquatic ecosystems as well as forest and non-forest ecosystems; and
  - providing background materials for improving the natural environment and landscape monitoring programs.

Denmark

Climate change impacts on water systems

Observed changes and trends	• Since the 19th century, the mean temperature has increased almost 1.5 °C. This increase is more than double the increase in the global mean temperature for the same period. • Annual precipitation has increased by about 15% (or 100 mm) since records began in 1874. The most significant increase in precipitation has occurred in west Jutland, where it has increased by about 20% in the past 85 years. • Fewer days with snow cover, decreased need for ice-breaking, shorter sledging season, earlier pollen season, longer growing season and longer swimming season, and so on.
Projected impacts	• Rise in annual mean temperature of 3 °C to 5 °C, depending on the emissions scenario. Warming leads to fewer days with frost and snow and fewer days with snow cover. • Increase of 10% to 40% in winter precipitation and a reduction in the order of 10% to 25% in summer precipitation. A clear tendency towards more episodes with very heavy precipitation particularly in autumn and lengthy dry periods especially in summer. • Overall, a positive effect on the runoff from land areas, with an increase of around 10% in the period December to April. More runoff in the Baltic region could make surface layers in the inner Danish waters less saline. • Decrease in average snow cover to about 25% of present-day values. • Reduced formation of groundwater in summer and an increased formation the rest of the year will affect the use of groundwater for drinking water, irrigation, etc. • Intrusion of saltwater may affect the quality of groundwater, which is significant because Danish water supply relies almost entirely on unpolluted groundwater. Greater risk of salt penetration may lead to limitations on water extraction. • Increase in the risk of flooding due to rising sea levels and increasing precipitation. • Extreme precipitation will generally become more extreme. The most extreme rainfall events will increase the most significantly. An increase in heavy downpours can be expected, especially during the summertime. Overall, heavy downpours will yield from 20% to 30% more water than today.

Primary concerns	Water quantity	Water quality	Water supply and sanitation	Extreme weather events	Ecosystems
			✓	✓	

Key vulnerabilities	• Increased demand for groundwater resources, possibly exacerbating the existing problem of over-use of groundwater resources (the main source of water supply) close to urban areas. • Significant increase in demand for irrigation, possibly exacerbating existing competition for water resources between agriculture and freshwater ecosystems.

Sources: Danish Meteological Institute (2011), *Adaptation to the Future Climate Change in Denmark – About the Information Centre for Climate Change Adaptation, http://en.klimatilpasning.dk* (accessed 20 September 2012); Ministry of Climate and Energy (2009), *Denmark's Fifth National Communication on Climate Change, http://unfccc.int/national_reports/annex_i_natcom/submitted_natcom/items/4903.php* (accessed 15 June 2012).

Key policy documents[1]

Document	Reference to water?	Type	Year	Responsible institution
Danish Strategy for Adaptation to a Changing Climate	Y	National adaptation strategy	2008	The Danish Government
How to manage cloudburst and rain water – Action plan for a climate-proof Denmark	Y	National adaptation action plan National risk assessment	2012 2012	Ministry of Environment Ministry of Environment

1. Climate change adaptation will be incorporated in the second planning period of the Water Framework Directive in 2015, based on the recommendation of a Danish advisory board on waterway legislation and the guidelines in White Book on Climate Change Adaptation by the EU Commission.

Policy instruments

Areas \ Policy mix	Regulatory instruments	Economic instruments	Information and other instruments
Water quantity			• The Information Centre for Climate Change Adaptation: A portal providing information on the latest research and developments on climate change adaptation in Denmark and abroad. The information is targeted at individuals, municipalities and businesses, *www.klimatilpasning.dk/EN-US/WATER/Sider/forside.aspx.* • Plan 09: Aims to facilitate adaptation measures in the municipal planning process by providing examples of adaptation and demonstrations projects from participating municipalities. • National Digital Elevation Model: Use to identify areas sensitive to future flooding.
Water quality			
Water supply and sanitation	• Reassessment of permits for water extraction: Possibility of relocation of water extraction in cases where it is not possible to maintain water levels in watercourses and wetlands while at the same time maintaining water supply (planned).		
Extreme weather events	• All 98 Danish municipalities are required to conduct a climate change adaptation plan before the end of 2013. This plan contains a risk-mapping of the entire surface area in each municipality for flood events from all water sources.[1] • Proposed legislation by the Environment Minister which would oblige all waste-water-companies to perform flood risk mapping of their sewer systems in relation to extreme rain water events (expected to take effect on the first of December 2012). • Law providing the municipalities with a legislative foundation for local city planning directly connected to climate change adaptation.[2] The new law allows municipalities to ban construction in certain areas solely due to climate change adaptation reasons.		
Ecosystems			

1. Flood risk mappings from rain events, sewers systems, creeks, sea and groundwater will be merged with a value-distribution-mapping to generate the local risk-map. The Danish government released the tools needed to conduct the mapping for the disposal of the municipalities in January 2013.
2. The law has passed by the Danish parliament on 29 May 2012.

Main research programmes

- Co-ordination Unit for Research in Climate Change Adaptation: Established by the Climate Change Adaptation Strategy, the unit will supply validated climate and climate-impact data, as well as specific research results of relevance to climate change adaptation, *www.klimatilpasning.dk/en-US/Service/Research/CoordinationUnitforResearchinClimateChangeAdaptation/Sider/Forside.aspx.*
- WaterCAP: Undertaken with 13 partners from Denmark, Belgium, The Netherlands, Germany and the UK, the project will provide recommendations for future implementation of EU directives and guidelines to take into account a change climate. It will generate policy relevant recommendations on: water quantity and quality issues; impact analysis of climate change on different types of water bodies; and adaptation strategy, including cross-sectoral involvement and a participatory approach, *www.klimatilpasning.dk/en-us/service/research/ongoingresearchprojects/sider/theclusterproject,watercap,willmakerecommendationsforeudirectivesandguidelines.aspx.*
- RiskChange: Statistically evaluating extreme precipitation, storm surges and extreme wave heights. It forms the basis of a risk assessment that is utilised in the design of infrastructure such as drainage systems, dykes and bridges, *www.klimatilpasning.dk/en-us/service/research/ongoingresearchprojects/sider/riskchange-risk-baseddesigninachangingclimate.aspx.*
- CLIWAT – Climate and Water: A Northern European project focussing on climate change impacts on groundwater. *www.klimatilpasning.dk/en-us/service/research/ongoingresearchprojects/sider/cliwat.aspx.*
- SWI: A strategic Danish Research Project about optimising integrated waste water systems, *www.klimatilpasning.dk/en-us/service/research/ongoingresearchprojects/sider/swi.aspx.*
- Water in cities: A new research partnership to help Danish cities adapt to the increased amounts of precipitation expected due to climate change, *www.klimatilpasning.dk/en-us/service/research/ongoingresearchprojects/sider/newpartnershiptosavecitiesfromflooding.aspx.*

Principal financing mechanisms and/ or investment programmes

- In 2012, the Danish Ministry of the Environment has supported 8 projects which show-case good examples of cross-sector co-operation on climate change adaptation with a total of EUR 360 000. Half of these projects relate to extreme rain events and the other half to the securing of coastline.
- Danish water and sewer companies will invest EUR 333.3 million (or 2.5 billion Danish kroner) in climate change adaptation in 2013 in the implementation of the agreement between the Danish government and the union of Danish municipalities for 2013, which details the budgetary limits of the 98 Danish municipalities.
- The Danish government is also currently scrutinising the Danish water sector legislation in order to prepare a new law proposal related to the financing of climate change adaptation of the water sector. The purpose of this work is to increase the possibilities for Danish water and sewer companies to rate finance more intelligent and socio-economically optimal climate change adaptation measures. For example, the proposal could make it possible for sewer companies to co-finance new measures on roads and in waterways, which keeps rainwater out of the sewer systems.

Highlights and innovative initiatives

- **Wetlands to reduce flood risks in Aarhus:** Water from heavy rainfall is being channeled to a recently restored wetland, Egå Engsø, provisioning flood protection for the low-lying and densely populated area near Aarhus, Denmark's second largest city. The wetland also reduces nitrogen leaching from surrounding agriculture. In 2007, following months of significant rainfall and a large amount of snowmelt, the limits of the flood protection systems were tested and required an emergency response to prevent a flood that could have had major economic consequences. This event highlighted the need for further preventive measures. A new wetland, H ede Enge, has been proposed to reduce risk from extreme rainfall events, which are projected to become more frequent and severe with climate change. The cost of the proposed project is estimated at approximately DKK 25 million, of which 80 per cent is for compensation for affected landowners for expropriation of land. Considered as a unique example of climate change adaptation, this project provides a good illustration of an eco-system based approach to adaptation for water systems, *http://klimatilpasning.dk/en-us/service/cases/sider/aarhusengelsk.aspx*.

Estonia

Climate change impacts on water systems

Observed changes and trends	• Annual mean temperature increased by 1.0 °C to 1.7 °C during the second half of the 20th century. • Since 1966, an increase in precipitation during the cold half of the year and also in June. A significant increase in precipitation (around 29%) has occurred in winter. • Significant decrease in the duration of snow cover and sea ice during the second half of the 20th century. • Apparent increasing trend in the inter-annual variability of the number of extreme wet, extreme dry and total number of extreme days over the past 50 years. Overall, a clear indication of a rising trend in extreme precipitation events during the period 1957-2006.				
Projected impacts	• The negative impacts of climate change in Estonia are expected to be less significant relative to other European countries. • Increase in groundwater recharge, depending on the hydro-geological conditions of catchments. Groundwater recharge is expected to be the most intensive in the Pandivere Upland, which is the most important groundwater catchment area in Estonia. The safe yield of wells abstracting from the upper aquifers will be augmented in Upper Estonia, which is expected to make the public water supply cheaper. • Earlier snowmelt causing shifts in the hydrological regime. Maximum river runoff will be reduced and will occur earlier. Lower water content of the soil and earlier appearance of drought conditions. • Drier climatic conditions in spring and in the first half of summer. • Shifting runoff regime, with decreases in spring runoff and increases in winter runoff, will have varied impacts on water resources management. More evenly distributed river flow will be beneficial for hydropower production. Increased flow in winter will improve water quality of rivers and benefits fish farm management. However, lower flows in the spring may deteriorate water quality and have a negative impact on aquatic habitats. • Increase in the temperature and the water balance of Lake Pepsi. Increase in water temperature result in an earlier and longer eutrophication period, impacting on water quality. • Increase in extreme weather events. Earlier and lower spring floods and more frequent winter floods. Shifting flood regimes may have an impact on infrastructures designed for past climate conditions with stable winters and higher spring floods.				
Primary concerns	Water quantity	Water quality	Water supply and sanitation	Extreme weather events	Ecosystems
	✓ (seasonal hydrological regime changes)	✓ (quality of groundwater)[1]	✓ (drinking water quality)	✓ (coastal and inland floods, extreme precipitation and temperatures)	✓ (increased sensitivity of ecosystems to human and climate pressures)[2]
Key vulnerabilities	• Drinking water quality degradation. • Eutrophication, with impacts on freshwater ecosystems. • Coastal areas, due to sea level rise and erosion. • The impact of winter and spring floods on inland water bodies, especially in densely populated areas.				

1. Deteriorating groundwater quality as heavy rains will cause increased leaching of pollutants into aquifers.
2. For instance, algae blooms will increase with the rise in water temperature.

Sources: European Climate Adaptation Platform (2012), *Estonia Countries Overview, http://climate-adapt.eea.europa.eu/countries/Estonia* (accessed 28 August 2012); Ministry of the Environment (2009), *Estonia's Fifth National Communication under the UNFCCC, http://unfccc.int/national_reports/annex_i_natcom/submitted_natcom/items/4903.php* (access 20 June 2012); Ministry of the Interior (2011), *Review of Emergency Risk Assessment, www.siseministeerium.ee/29960.* (accessed 28 August 2012).

Key policy documents

Document	Reference to water?	Type	Year	Responsible institution
Emergency Act and Emergency Risk Assessment[1]	Y	Legal act	2009/2011	Ministry of the Interior
Water Act[2]	Y	Legal act	2009	Ministry of the Environment
National Adaptation Strategy		National adaptation strategy	Under development, planned for 2016	Ministry of the Environment
Nature Conservation Development Plan up to 2020	Y	National strategy	2012-20	Ministry of the Environment
National adaptation plan		National adaptation plan	Under development, planned for 2016	Ministry of the Environment
River basin management plans and flood risk management plans	Y	River basin adaptation plans	Under development, planned for 2015	Ministry of the Environment
National vulnerability assessment		National vulnerability assessment	Under development, planned for 2016	Ministry of the Environment
Various projects in the Baltic Sea[3]	Y	Transboundary responses	Under development	

1. Entered into force on 24 July 2009, the Emergency Law requires the establishment of risk assessments and crisis management plans in case of storms and floods at least once every two years. The first Emergency Risk Assessment was compiled in 2011. These plans are prepared in co-operation between different institutions, ensuring better communications and clarity of roles and responsibilities, *www.siseministeerium.ee/29960*.
2. Estonia adopted the requirements of the EU Directive 2007/60/EC on the assessment and management of the flood risk in November 2009.
3. Various projects with a transboundary component include: BaltAdapt, *www.baltadapt.eu*; BaltClim, *www.bef-de.org/unsere-themen-en/projects/baltclim*; BaltCICA, *www.baltcica.org*.

Policy instruments

Areas \ Policy mix	Regulatory instruments	Economic instruments	Information and other instruments
Water quantity	• Water Act along with the implementing acts.	• Pollution charges, fines.	• The Emergency Act: Requires emergency risk assessments and crisis management plans in the case of storms and floods. • Rescue Centres: Improved crisis communication for extreme weather events. In 2009, a nation-wide radio communication system was implemented to facilitate information exchange between agencies. • Risk analysis of extreme events by cities: Local plans take into account new flood risks.
Water quality	• Water Act along with the implementing acts. • Public Health Act, along with the implementing acts.	• Pollution charges, fines.	
Water supply and sanitation	• Water Act along with the implementing acts. • Public Water Supply and Sewerage Act, along with the implementing acts.	• Pollution charges, fines.	
Extreme weather events	• Water Act, along with the implementing acts. • Rescue Act, along with the implementing acts. • Emergency Act, along with the implementing acts.		
Ecosystems	• Water Act, along with implementing acts.	• Fines.	

Main research programmes

- Meteorological and Hydrological Institute of Estonia: Provides data and climate information on weather observations and scenarios, weather events and climate change science, *www.emhi.ee/index.php?nlan=eng*.
- On-line Sea Level Information System hosted by the Marine Systems Institute at Tallinn University of Technology provides information about sea level status, trends, projections, and water temperatures in different coastal regions of Estonia. National monitoring program (monitoring different water, air quality and biodiversity parameters), *http://on-line.msi.ttu.ee/kaart.php*.

Principal financing mechanisms and investment programmes

- Environmental Investment Centre provides grants for various activities concerning water management and climate change, drawn from two sources of financing – the Cohesion Fund and the Environmental Programme (the ambient air protection, water and nature protection programmes).

Finland

Climate change impacts on water systems

<table>
<tr><td>Observed changes and trends</td><td colspan="5">• Increase in annual mean temperature by approximately 1 °C since the middle of the 19th century. Warming has been the most intense in the winter months. Since the 1970s, temperatures have rapidly, particularly in winter.</td></tr>
<tr><td>Projected impacts</td><td colspan="5">• Increase in annual mean temperature increase of 3 °C to 6 °C for the period 2070-99, as compared to 1971-2000. Greater increase in winter than in summer. Temperature increase in Finland is expected to be about 1.5 times higher than the global average temperature rise.
• Increase in precipitation likely to be substantial. Increase in annual precipitation of 10% to 25% in the period 2070-99, as compared to 1971-2000. Greater proportional increase in winter than in summer. Decrease in the ratio of snow to rain.
• Shift in the seasonal distribution of runoff is considered to be the most important climate change impact on hydrological regime. Winter runoff is expected to increase due to an increase in snowmelt and rainfall. Changes in yearly runoff are estimated to be between -5% and +10%, depending on the catchment area and its lake's baseline percentage.
• Shorter and discontinuous snow season. Decrease in snow cover progressing from south to north. In the North, an increase of snow in the near-term due to increased snowfall.
• Longer dry period in summer in southern Finland will reduce groundwater (high quality water for households) quality and discharge. Changes in rain conditions will affect the quality of water and increase the pressures on water systems (e.g. nutrient and organic loads) and may deteriorate groundwater quality.
• Increased precipitation and more even discharge (smaller spring floods and larger discharge in winter) will be beneficial for hydropower production. It is likely that additional capacity will be built alongside existing hydropower plants.
• More frequent winter floods and a smaller amount of snow (reducing natural storage in snow pack) will mean greater storage capacity requirements in reservoirs and winter flooding in large central lakes, such as Saimaa and Päijänne.
• Greater intensity and frequency of heavy rainfall. Increase in intense rainfall of up to 40% to 60% expected to cause problems for dam safety, particularly in small rivers and for dams in large rivers with older flood design estimation. Increase in risks related to water works and discharges from stormwater.
• Decrease in spring floods in southern and central Finland. Mean summer rainfall might decrease, but more frequent and severe summer floods are expected due to increased extreme rainfall.
• Higher temperature and longer summer periods with higher evaporation could cause intense and prolonged drought periods.
• Increasing temperatures and runoff and the resulting changes in nutrient loads may have a profound impact on aquatic ecosystems e.g. phytoplankton, fish stock, etc. Some species characteristic to Finland, like relict cold water fish may become extinct, even if climate change will probably increase the total number of Finnish flora and fauna.</td></tr>
<tr><td rowspan="2">Primary concerns</td><td>Water quantity</td><td>Water quality</td><td>Water supply and sanitation</td><td>Extreme weather events</td><td>Ecosystems</td></tr>
<tr><td></td><td></td><td></td><td>✓</td><td></td></tr>
<tr><td>Key vulnerabilities</td><td colspan="5">• Small water utilities and wastewater systems with combined sewers are vulnerable to climate-related problems.
• Vulnerability of housing areas and infrastructure may increase with possible increase in hydro-meteorological extremes.</td></tr>
</table>

Sources: Finnish Environment Institute (2011), *Vulnerability Assessment of Ecosystem Services for Climate Change Impacts and Adaptation: Key Results*, *www.syke.fi/en-US/Services__Data/Research_and_development_projects/Projects/Vulnerability_Assessment_of_ecosystem_services_for_Climate_Change_Impacts_and_Adaptation_VACCIA/Vulnerability_Assessment_of_ecosystem_se%2810101%29* (accessed 12 August 2012); Ministry of Agriculture and Forestry (2005) *National Strategy for Adaptation to Climate Change*, *www.mmm.fi/en/index/frontpage/climate_change_energy/adaption.html* (accessed 12 August 2012); Ministry of the Environment (2009), *Finland's Fifth National Communication under the UNFCCC*, *http://unfccc.int/national_reports/annex_i_natcom/submitted_natcom/items/4903.php* (access 20 June 2012).

Key policy documents

Document	Reference to water?	Type	Year	Responsible institution
National Strategy for Adaptation to Climate Change	Y	National adaptation strategy	2005	Ministry of Agriculture and Forestry
Action Plan for the Adaptation to Climate Change	Y	National adaptation plan	2008 (updated in 2011)	Ministry of Agriculture and Forestry, Ministry of the Environment
Map-based assessment of vulnerability to climate change employing regional indicators (MAVERIC)	Y	National vulnerability assessment	2009-11	
Vulnerability Assessment of ecosystem services for Climate Change Impacts and Adaptation (VACCIA)	Y	National vulnerability assessment	2009-11	Finnish Environment Institute (SYKE)

Policy instruments

Areas \ Policy mix	Regulatory instruments	Economic instruments	Information and other instruments
Water quantity	• Review of existing permits for each catchment: Existing arrangements will be examined to determine how effective they are in responding to changing water conditions. If needed, measures will be taken to enhance the flexibility of permits.		• Pilot project on preparing for climate change in regional and general planning: undertaken by the Ministry of the Environment/Department of the Built Environment as part of promoting the implementation of the national land use guidelines. • Climate change communications network: launched in 2008 by the Ministry of the Environment and the Ministry of Agriculture and Forestry, aims to promote communication and co-operation between various branches of administration and research institutes. • Water resources management plans and flood risk management plans will be analysed for their climate impact and resilience to climate change. • Flood forecasting and monitoring: Flood warning by the Finnish Environment Institute will be further improved as part of the national early warning systems for natural disasters, *www.environment.fi/waterforecast*. • Stormwater management manual: Greater use of this manual will be promoted.
Water quality	• In the review of River Basin Management Plans the impacts of climate change on water quantity and quality will be assessed and measures will be checked for climate proofing.		
Water supply and sanitation	• Development of instructions for the preparedness of water treatment plants for special weather conditions. • Revision of Water Supply Act to improve storm water management (2012/2013).		
Extreme weather events	• Revision of Dam Safety Act (2008) including periodic update of safety checks and reports. • Review of the Land Use and Building Act and Decree: Will be undertaken to examine whether the Act needs to be amended with regard to safeguarding already built-up areas against floods.	• Reform of the compensation system for flood damages to better account for a changing climate and respond to extreme weather conditions.[1]	
Ecosystems			

1. The compensation system for flood damages will apply uniform compensation for flood damage to buildings and their contents, regardless of the cause of flooding.

Main research programmes

- FINADAPT: Under the Environmental Cluster research programme, assessed the level of knowledge, gaps in knowledge and research needs concerning adaptation in Finnish nature and society, *www.environment.fi/default.asp?contentid=365716&lan=EN*.
- Climate Change Adaptation Research Programme (ISTO): Was launched as part of the implementation of the National Strategy for Adaptation to Climate Change. It aims to produce information that will facilitate the planning of practical adaptation measures. Over the period 2006-10 funding totalling EUR 0.5 million was allocated to 30 research projects, *www.mmm.fi/en/index/frontpage.html*. ACCLIM scenario and information service project is a part of this programme.
- Finnish research programme on climate change (FICCA, 2011-14): Launched to respond to the scientific challenges posed by climate change on a broad front, *www.aka.fi/ficca-en*.

Principal financing mechanisms and investment programmes

France

Climate change impacts on water systems

Observed changes and trends	• From the end of the 1980s, average temperatures have risen rapidly, faster than the global average. Overseas territories show a similar trend. • The French Alpine glaciers have shrunk and the glaciers in the Pyrénées are showing the same trend. • The fauna and flora are also affected. The influence of climate change on vine maturity dates and thus the dates for the grape harvest is well known, but fruit and timber trees are also seeing a variation in their growth cycle.
Projected impacts	Projections to the year 2050 include: • Moderate reduction in the overall average of flows in winter, except for the Southeast and the Alps, where they will increase. In spring, slight changes, in general. In summer and autumn, a major reduction in flows. • Significant increase in the number of days with low water level. • Reduction in soil humidity regardless of the season, except in mountain areas in winter and/or in spring. • Sharp decrease in snow precipitation and maximum height of accumulated snow at low altitude, an effect which lessens as the altitude increases. • Under the assumption of stable individual demand, an increase of 2 billion m^3 in water supply is anticipated to be required in order to meet the current needs of industry, agriculture and households. • Reduction in the quantity of water resources, coupled with a potential increase in anthropogenic pressure due to demographic growth, could have significant impacts on water quality. • Reduction in flood flows well below average in some cases, but an increase in other cases. • No clear trend towards a sharp increase in the risk of flooding by overbank flow, given the current state of knowledge. • Increase in the duration and intensity of summer droughts is significant in all regions.

Primary concerns	Water quantity	Water quality	Water supply and sanitation	Extreme weather events	Ecosystems
	✓ (additional stress in the regions already adversely affected)			✓ (drought in the south of France)	

Key vulnerabilities	• Irrigation in the southwest in the future.

Sources: Observatoire national sur les effets du réchauffement climatique (2009), *Climate Change: Costs of Impacts and Lines of Adaptation*, *www.developpement-durable.gouv.fr/ONERC-Report-to-the-Prime-Minister.html* (accessed 29 June 2012); Ministère de l'Écologie, du Développement durable et de l'Énergie (2011), *Plan national d'adaptation de la France aux effets du changement climatique 2011-2015*, *www.developpement-durable.gouv.fr* (accessed 29 June 2012) ; Ministère de l'Écologie, du Développement durable et de l'Énergie (2009), *The Fifth National Communication of France to the UNFCCC*, *http://unfccc.int/national_reports/annex_i_natcom/submitted_natcom/items/4903.php* (accessed 20 June 2012); Ministère de l'Écologie, du Développement durable et de l'Énergie (2006), *Stratégie nationale d'adaptation au changement climatique*, Observatoire National sur les Effets du Réchauffement Climatique, *www.developpement-durable.gouv.fr* (accessed 29 June 2012).

Key policy documents

Document	Reference to water?	Type	Year	Responsible institution
Water and Aquatic Ecosystems Law (La Loi sur l'eau et les milieux aquatiques – LEMA)	Y	Legal act	2006	Ministère de l'Écologie, du Développement durable et de l'Énergie (MEDDE)
National Strategy for Climate Change Adaptation (Stratégie nationale d'adaptation au changement climatique)	Y	National adaptation strategy	2006	MEDDE
Climate Change: Costs of impacts and lines of adaptation	Y	National risk assessment	2009	Observatoire national sur les effets du réchauffement climatique (ONERC)

Key policy documents *(cont.)*

Document	Reference to water?	Type	Year	Responsible institution
National Climate Change Adaptation Plan (Plan national d'adaptation au changement climatique)	Y	National adaptation plan	2011	MEDDE
Regional Climate, Air and Energy Programme (SRCAE) and Regional Climate-Energy Plans (PCET) has regional responsibilities for adaptation		Sub-national responses (legally-binding)	Under development	State and Regions (SRCAE) Local community (PCET)
Management plans (*Plans de gestion*) composed of water planning and management plans (*Schémas directeurs d'aménagement et de gestion des eaux* – SDAGE) and of measures' programs (Programmes de mesures)	Y	Basin level planning documents	2010-15	Water Agencies

Policy instruments

Areas \ Policy mix	Regulatory instruments	Economic instruments	Information and other instruments
Water quantity	• New procedure for defining abstractable volumes: in sub-basin areas, defining abstractable volumes by each kind of water users (households, farmers and industries), while maintaining a minimum ecological flow, in order to ensure that the demand does not exceed the local supply. These volumes will be updated every few years, so that the updated volumes match with the trend of the supply changes, including climate-related impacts.	• Regular update of the water environmental taxes' rates in the Water Agencies' Intervention Programmes (every 5 five years), in order to meet the new water challenges, including those related to the decrease in water availability.[1]	• Communication portal to diffusion information on climate change impacts. • Target to save 20% of water withdrawn (excluding water storage in water) by 2020, compared to the yearly average of the period 2005-10. Measures include: Communication about the benefits of saving water as well as financial support for water-saving activities and promotion of rainwater collection. • Information portal on drought and water restrictions (PROPLUVIA): Provides real-time, searchable database and maps of possible water restrictions by department, *http://propluvia.developpement-durable.gouv.fr/propluvia/faces/index.jsp* (in French). • Flood risk mapping for regions at high risk of flood will be included in each risk management plan. • Guidance on integrating climate change into the planning process for water management (MEDDE, Water Agencies). • Mapping of vulnerability of groundwater to climate change (ONEMA). • Identification of possible scenarios for adaptation of water-consuming activities in areas already suffering from water scarcity (MEDDE, with others). • Information portal on major risks: provides easily accessible information on major risks (including natural disasters, such as floods) for the general public, *www.risques.gouv.fr* (in French).
Water quality	• See above regarding the minimum ecological flows that contribute to maintaining good ecological status of aquatic environments.		
Water supply and sanitation	• Regulatory framework and financial support to promote wastewater reuse for irrigation of crops and green spaces.	• Network taxes for the users connected to the water supply and/or sanitation networks, in order to contribute to an appropriate renewal of the water infrastructures and devices.[2]	

Policy instruments *(cont.)*

Areas \ Policy mix	Regulatory instruments	Economic instruments	Information and other instruments
Extreme weather events	• Acceleration of the plan to prevent flood risk: Aims to limit impact of floods by restricting the construction of buildings in risk-prone areas and by protecting construction areas from flooding, as well as stipulating measures to reinforce existing buildings, *www.developpement-durable.gouv.fr/L-acceleration-des-Plans-de.html* (in French). • National Flash Flood Plan[3] (Plan Submersion Rapide, PSR): Proposes actions to ensure the safety of people facing several types of flood hazards. The plan comprises national actions (regulations, information-based instruments) shared across levels of government, as well as local projects. • Integration of the impacts of climate change on natural risks into urban planning process: In the implementation of the Flood Directive, climate change adaptation will be included as a strategic component in local strategies.	• Reform of the insurance scheme for natural catastrophes[4] (CatNat): Undergoing study concerning the possible adjustment of insurance rates to support increased responsibility of actors regarding their actual risk exposure. A law proposal to enable rates adjustment to the level of investment in risk reduction for firms or local collectivities is currently under review at the Senate. • Promote increased subscription to home insurance in overseas territories, so that households will benefit from CatNat coverage. • "Rain tax" (Décret, 2011): Allows municipalities to institute a tax to manage urban rain water by providing an incentive to limit impervious surfaces and promoting infiltration.	
Ecosystems	• Minimum ecological flows contribute to the maintaining of good ecological status of aquatic environments (see above description on water quantity).	• Water Agencies' subsidies to projects aiming to protect and restore environments damaged by water deficit, among other environmental issues.	

1. The new rates are proposed by the Water Agencies, and then the Basins' Committees (composed of users, State and Water Agencies) discuss these rates and vote on the final ones.
2. For the rates' sizing, see above the explanation for the water environmental taxes.
3. Released in July 2012.
4. The French insurance system for natural catastrophes (CatNat) is based on the principle of solidarity in three main ways: *i)* by the legal obligation to have insurance since property insurance obligatorily covers natural disasters; *ii)* by a uniform rate for the extra CatNat premium paid by any policy-holder (12% for a package policy for dwellings, 6% for an insurance contract for vehicles); *iii)* by the State guarantee to the Central Reinsurance Fund. 60% of the CatNat compensations over the 1982-2006 period (EUR 7.3 billion) concerned damage from floods. Bommelaer et al.(2011), *Financing Water Resources Management in France – A Case Study for an OECD Report*, Studies and documents of the Department for the Economics, Assessment and Integration of Sustainable Development.

Main research programmes

- Explore 2070: Developed a systemic vision of the impacts linked to water cycle changes, the costs and risks associated with different climate change scenarios and examine possibly adaptation strategies for mainland France and overseas departments from now to 2070 (MEDDE/DGALN/DEB).
- Project to develop new knowledge at the scale of large river basins, based on hydrological modelling incorporating the impacts of climate change (Water Agencies).
- Project to evaluate the impact of climate variability of low-flow regimes (ONEMA).
- Study to evaluate conditions to implement active groundwater management (ONEMA/DBSN/MTES).
- Project on consolidating knowledge of flood risks and assess the impact of climate change (MEDDE/DGPR).
- Inventory of flood prevention measures and development of a decision support tool (MEDDE).
- Project to continue mapping flood risks for high-risk areas (in the context of the implementation of the EU Flood Directive).

Principal financing mechanisms and investment programmes

- The National Adaptation Plan (2011-15): Enumerates 84 actions and 230 specific measures at a cost of approximately EUR 171 million. In addition to this sum, directly dedicated for new measures, an additional EUR 391 (*Investissement d'avenir*) will be invested in research and innovation that will contribute in some way to adaptation. In addition, EUR 500 million will be invested over the period 2011-16 to implement the Drought Plan (*Plan sécheresse*) and the National Flood Plan (*le Plan submersions rapides*).
- Introduction of eligibility criteria to avoid mal-adaptation for relevant public and private funding mechanisms: Initiative to include climate change adaptation as eligibility criteria for financing of long-lived investments. A methodological approach and tool have been developed and will be published in autumn 2012. These include a water consumption trend component. Mobilisation of financial resources for adaptation: Identification and review of how existing financing mechanisms could potentially be modified to support adaptation as well as a study of potential sources of additional financing. Specific examples include the update of the abstraction tax rates every 5 years, in every basin, in order to: *i)* improve the tax revenues; and *ii)* send an appropriate price signal, reflecting local scarcity, to the user. This also includes Water Agencies' subsidies to water environmentally-friendly projects. These economic instruments follow the objectives defined in the SDAGE (see table above), which now include climate change components.
- Study of potential innovative financing mechanisms to provide incentives for individuals to adapt to climate change. For example, the new rainwater management tax (only in communes that choose to implement it), in order to discourage soil sealing and thus to contribute to natural groundwater recharging.
- Several investment programmes for Water Agencies to reduce water loss and augment supply: specifically aims at reducing water leakage within drinking water networks, supporting wastewater reuse and optimising water storage and development of new water storage, as needed. There are subsidies for environmentally-friendly water-related projects.[1]
- National investment in Innovation and Research, Investment in the Future Programme (Investissement d'avenir): Will scale up adaptation research.

1. The amounts are available for each Water Agency in the "Projet de Loi de Finances 2012" (Tax revenues and expenses' reports that are voted by the Parliament) are available at: *www.performance-publique.budget.gouv.fr/farandole/2012/pap/pdf/Jaune2012_agences_eau.pdf* (in French) EUR 12 millions in 2010 (pp. 22-23).

Highlights and innovative initiatives

- **Reform of the Insurance scheme for Natural Disasters (CatNat):** Ongoing reforms on the suitability of the systems in the context of climate change. Reforms aim at making insurance an effective economic incentive to risk prevention (see above reference to the proposal in the Senate).
- **Promotion of water use efficiency:** By 2020, target to save 20% of water withdrawn (excluding water storage in winter), compared to the yearly average of the period 2005-10. Measures to achieve this saving include: Communication about the benefits of saving water as well as financial support for water-saving activities and promotion of rainwater collection (see above the reducing water leakage projects, but also, promoting rainwater collection). It is possible to finance these devices thanks to tax credits known as "Crédits d'impôts développement durable" (sustainable development tax credits).
- **Plan to prevent flood risks:** Aims to limit the impact on human lives, damage to buildings and commercial activities, and natural hazards, mainly by restricting the construction of buildings in risk-prone areas and by protecting construction areas from flooding, as well as stipulating measures to reinforce existing buildings. In 2013, all risk-prone areas should be covered by the plan, *www.developpement-durable.gouv.fr/L-acceleration-des-Plans-de.html* (in French).

Germany

Climate change impacts on water systems

Observed changes and trends	• Increase in the regional average for mean annual precipitation by about 9%, compared with the beginning of the 20th century. • Change in the distribution of rainfall over the summer months; reduced rainfall in July and August is largely offset by heavier rainfall in June. • Winter rainfall shows a general increase of around 20%, but the increase of winter rainfall in eastern parts of Germany is compensated by the decreased rainfall in summer.
Projected impacts	• Increase in the probability of flooding as a result of heavier and more frequent intense rainfall. • Milder winters will reduce the proportion of total precipitation accounted for by snow. This will reduce temporary storage of water in the form of snow, which means that precipitation will result in immediate runoff. The danger of winter flooding will increase. • More frequent occurrence of low-water periods, usually as result of prolonged dry periods in summer, with effects on water abstraction for cooling and aquatic ecosystems. In addition, earlier melting of snow will mean reduced compensation of low water in the Rhine and Danube during the summer months. • Possible increase in extreme wind and rainfall events increases the danger of erosion, and this may result in pollutants, fertilisers and pesticides from a variety of areas entering the groundwater and surface waters. • During intense rainfall events the capacity of combined sewage systems in settlement areas may be exceeded, thereby resulting in increased nutrient inputs into rivers and lakes. As a result, there may be extremely frequent but localised appearances of pathogens. • Increased flood events and low summer water levels in surface waters increase the concentration of undesirable substances in the water. Although the input of hazardous substances caused by floods is completely different from the rising concentrations during low water situations, in general, both climatic phenomena may place a strain on ecosystems with regard to substances. In cases of drinking water abstraction from bank-filtered water, the complexity and cost of drinking water purification may increase. • A progressive rise in water and soil temperatures of aquatic systems in the summer can lead, for example, to a reduction in the oxygen concentration of lakes and rivers. For aquatic fauna and flora, this means additional stress, since they are already suffering from high water temperature and limited water volume. Low oxygen concentrations and high water temperatures favour re-dissolution from sediments during low-water periods, and may therefore result in entrainments of nutrients into the water. In addition, the worsening of the dilution ratio leads to increasing pollution of waters by discharges. • Since drinking water supplies in Germany mostly make use of local groundwater resources and are only drawn to a limited extent from bank-filtered water or surface water (e.g. reservoirs), no fundamental drinking water supply problems are expected even under changed climatic conditions. Nevertheless, regional exceptions cannot be ruled out in areas suffering from prolonged droughts.

Primary concerns	Water quantity	Water quality	Water supply and sanitation	Extreme weather events	Ecosystems
	✓ (with regionally variation)			✓	

Key vulnerabilities	• Seasonal drought risks in eastern Germany with potential impacts on agriculture and forestry. • Flooding risks in some river catchments (Rhine, Danube, Elbe, Oder).

Sources: Federal Republic of Germany (2010), *Fifth National Report of the Government of the Federal Republic of Germany under the UNFCCC, http://unfccc.int/national_reports/annex_i_natcom/submitted_natcom/items/4903.php* (accessed 20 June 2012); Federal Republic of Germany (2008), *German Strategy for Adaptation to Climate Change*, Strategy adopted by the German Federal Cabinet.

Key policy documents

Document	Reference to water?	Type	Year	Responsible institution
Federal Water Resources Act (including flood risk management)	Y	Legal act	Latest amendment in 2010	Federal Government
Federal Regional Planning Act	Y	Legal act	Revised in 2008	Federal Government
German Strategy for Adaptation to Climate Change (DAS)	Y	National adaptation strategy	2008	Federal Ministry for the Environment, Nature Conservation and Nuclear Safety (BMU)

Key policy documents (*cont.*)

Document	Reference to water?	Type	Year	Responsible institution
Adaptation Action Plan of the German Strategy for Adaptation to Climate Change (APA)	Y	National adaptation plan	2011	BMU
Climate Change in Germany – Vulnerability and Adaptation of Climate-Sensitive Sectors Vulnerability Network	Y	National risk assessment	2005	Federal Environmental Agency
Länder Adaptation Strategies Individual cities and local authorities	Y	Sub-national responses	Under development	Individual *Länders* Individual cities and local authorities

Policy instruments

<table>
<tr><th>Areas \ Policy mix</th><th>Regulatory instruments</th><th>Economic instruments</th><th>Information and other instruments</th></tr>
<tr><td>Water quantity</td><td>• Review of the Federal Water Resources Act: The Act (entered into force in 2010), was reviewed to implement the EU-Flood Risk Management Directive and to examine whether it is necessary or appropriate to include climate change impacts or adaptation requirements as a target, principle or trade-off.</td><td>• In 13 out of 16 Federal States, there are laws on water abstraction fees.[1]</td><td rowspan="5">• KomPass (Climate Impacts and Adaptation in Water Management): The national information, communication and co-operation platform on adaptation, www.anpassung.net/cln_108/nn_700470/DE/Fachinformationen/KlimaFolgen Anpassung/Wasserwirtschaft/wasser wirtschaft__node.html?__nnn=true (in German).
• The "Deutscher Wetterdienst" presents information about current weather situation in Germany, official warnings for severe weather events and information about the current climate and the climate in the future www.dwd.de.
• Climate Service Centre refines the knowledge derived from climate research in a practice-orientated way and conveying the findings to decision-makers in politics, administration, and economy and for the broad public, www.climate-service-center.de.
• Platform on Flood Risk Information and Warning, www.hochwasserzentralen.de (in German).
• Information about water levels of the Federal Water Ways, www.pegelonline.wsv.de/gast/start (in German).</td></tr>
<tr><td>Water quality</td><td>• Ordinance for the Protection of Surface Waters of 20 July 2011: Implements the environmental quality standards of the EU Water Framework Directive.
• Ordinance for the Protection of Groundwater of 9 November 2010: implements the environmental quality standards of the EU Groundwater Directive.</td><td>• Waste Water Charges Act (latest amendment 2005): Regulates the levying of charges for the direct discharge of waste water into water bodies.</td></tr>
<tr><td>Water supply and sanitation</td><td>• Drinking Water Ordinance of 2011: Implements the quality standards for drinking water of the EU Drinking Water Directive.</td><td>• The Acts on municipal charges of the Federal States determine that the principle of full cost recovery is applied for charges paid by the consumers for water supply and waste water discharge.</td></tr>
<tr><td>Extreme weather events</td><td>• Revision of the Federal Water Resources Act in 2010 to implement the EU Flood Risk Management Directive.
• Revision of the Federal Regional Planning Act: In 2008, adaptation to climate change was introduced into this legislation as one of the principles of spatial planning.[2]
• Revision of physical planning law: In June 2011, the decision was taken to emphasise climate-friendly urban development (climate protection and adaptation to climate change) as a guiding principle for the planning process.</td><td></td></tr>
<tr><td>Ecosystems</td><td></td><td></td></tr>
</table>

1. See pages 107 ff in *www.umweltdaten.de/publikationen/fpdf-l/4189.pdf* (in German).
2. This put in place a framework that will allow the spatial plans of the *Länder* and regions to be gradually supplemented with the aspect of provision for the spatial requirements of climate adaptation during their redrafting process. In every revision of spatial plans, sectoral environmental plans will be incorporated.

Main research programmes

- National Meteorological Service: Provides climate monitoring, including the provision of climate projections for the planning and preparation of adaptation measures, *www.dwd.de/bvbw/appmanager/bvbw/dwdwwwDesktop?_nfpb=true&_windowLabel=dwdwww_main_book&T175 00013121287044130563gsbDocumentPath=&switchLang=en&_pageLabel=dwdwww_start*.
- Germany's Federal Institute of Hydrology (BfG) and the Federal Maritime and Hydrographic Agency (BSH): Collect data and develop climate projections for inshore waters. BfG: *www.bafg.de/cln_031/nn_161876/sid_53259BD539D84CE3CD4572BF6310A75E/nsc_true/EN/Home/homepage__en__node.html?__nnn=true*; BSH: *www.bsh.de/en/index.jsp*.
- KLIWAS Impacts of Climate Change on Waterways and Navigation: A consortium research programme by the Federal Ministry of Transport (BMVBS) to survey adaptation needs on the basis of the findings and draw up options for the adaptation measures that will be required, *www.kliwas.de/cln_033/nn_523302/KLIWAS/EN/Home/homepage__node.html?__nnn=true*.

Principal financing mechanisms and investment programmes

- The Adaptation Action Plan will be funded from the budgets of the respective governments' departments within the current financial planning.
- The Federal Government will examine the inclusion of aspects of climate adaptation in Federal funding programmes relevant to adaptation, and will also examine joint funding instruments financed by the Federal Government, the Lander and the EU. The recent incorporation of adaptation into the funding instruments of the National Climate Protection Initiative is an example.
- At the end of 2011, BMU introduced a funding scheme promoting adaption to climate change at the level of individual enterprises and local authorities. This funding is expected to cover networking and education projects at the local/regional levels and support for drawing up adaptation concepts. Funding is also available under the National Climate Initiative for municipalities for developing climate change adaptation concepts as part of municipal climate strategies.
- The Federal Government is funding exemplary models and demonstration schemes at local and regional level, such as "Managing Climate Change in Regions for the Future (KLIMZUG)", "Urban strategies to Combat Climate Change", etc. This scheme provides financial incentives for frontrunners to foster innovation and to spread awareness about the necessity of adaptation.

Highlights and innovative initiatives

- Because a national insurance duty for natural hazards does not exist in Germany, there are **awareness raising campaigns** on *Länder* level with the goal to raise the percentage of people and enterprises with voluntary insurances against natural hazards.

Greece

Climate change impacts on water systems

Observed changes and trends	• Annual increase in temperature of about 0.4 °C to 0.6 °C since the 1990s, as compared to the mean values of 1961 to 1990. The increase is mostly due to a steady rise of temperature during the summer period (from April to September). • Significant reduction of the precipitation in recent years, especially in the second part of the 20th century. • Significant increase in the frequency of extreme events in the last two decades. Heat waves have occurred every year since 1997. Major damaging floods have also occurred in the recent years.
Projected impacts	• Significant increase in annual maximum temperature compared to the northern European countries. • Decrease in mean annual precipitation. Change in seasonal precipitation varies substantially from season to season and across regions. Decrease in winter precipitation and a substantial decrease in summer precipitation. Despite the decrease in mean precipitation, the intensity of precipitation is protected to increase. • Quantity and quality water problems caused by climate change are mainly attributed to the projected decrease of precipitation during the summer months and also the salination of the water as a result of sea level rise. • Major drought episodes are projected to become more frequent with particularly intense summer droughts. Greece is among the countries expected to be the worst hit by an increase in frequency and severity of droughts and water scarcity.

Primary concerns	Water quantity	Water quality	Water supply and sanitation	Extreme weather events	Ecosystems
	✓ (shortage)			✓ (droughts)	

Key vulnerabilities	• In the Mornos River Basin the pressure on the water supply system of the city of Athens has been a very significant issue.

Source: Ministry of Environment, Energy and Climate Change (2010) *Fifth National Communication to the UNFCCC on Climate Change, http://unfccc.int/national_reports/annex_i_natcom/submitted_natcom/items/4903.php* (accessed 20 June 2012).

Key policy documents

Document	Reference to water?	Type	Year	Responsible institution
Water Law[1]	Y	Legal act	2003	
Water Scarcity and Drought Management Strategies	Y	National strategies related to water		Ministry of Environment, Energy and Climate Change (MEECC)
National Strategy for Adaptation to Climate Change		National adaptation strategy	planned	MEECC

1. Law 3199/2003, adopted in December 2003.

Policy instruments[1]

Areas \ Policy mix	Regulatory instruments	Economic instruments	Information and other instruments
Water quantity			• Civil Protection planning and actions for flood risk: A circular explicitly setting out the obligations of public authorities before, during and after a flood for the whole of the Greek Territory was issues in 2009 by the General Secretariat of Civil Protection, *www.gscp.gr/ggpp/site/home/ws.csp?loc=en_US*.
Water quality			
Water supply and sanitation		• Rational pricing policies for municipal water services: Progress is being made towards full cost recovery from household billing, which currently amounts to about 60% on average.[2]	
Extreme weather events	• Uneven spatial and seasonal distribution of water resources in Greece is being addressed more effectively than in the past, aiming at adapting to prolonged droughts.		
Ecosystems			

1. The adaptation measures that are currently under implementation in Greece are part of a broader network of measures that applies to the specific areas of identified vulnerabilities.
2. According to the Central Water Agency of the MEECC, there is quite a variation among basins, with Attica and Thrace showing a cost recovery of 108% and 103%, respectively, whereas Thessaly and East Peloponnese only reach 34% and 38%, respectively.

Main research programmes

- Research group "Atmospheric Chemistry and Climate Change Modelling" of the National Observatory of Athens has undertaken various modelling simulations of the future climate, *www.meteo.noa.gr*.

Principal financing mechanisms and investment programmes

- Acheloos water transfer project includes the construction of four major dams and reservoirs, a 17.4 km long diversion channel to Thessaly and two tunnels.

Hungary

Climate change impacts on water systems

<table>
<tr><td>Observed impacts</td><td colspan="5">• In 2006, there were two extreme floods in the Tisza and in the Danube that occurred at the same time. Due to the extreme weather conditions of the past years, there were serious inland damage from excess water and other floods.
• Following a period of dryness and drought lasting nearly two decades, four highly dangerous floods occurred in the Hungarian part of the River Tisza from 1998 to 2001.</td></tr>
<tr><td>Projected impacts</td><td colspan="5">• Significant increase in temperature is highly probable in all seasons as well as increase of approximately 1 °C to 1.4 °C in the annual average.
• Changes in annual and seasonal precipitation are not significant at the national level. However, the regional and territorial changes might be significant. Precipitation may either increase or decrease.
• Increase in summer temperature and decrease in summer precipitation will result in less drainage in summer, decrease in the water reserves of lakes in summer, decreasing humidity of soil, and longer dry periods.
• Increase in winter temperature, change in the amount of winter precipitation and an increase in the proportion that will fall as rain will result in reduced snow cover. This will also result in rain related drainage increases in winter, earlier floods, rivers peak at higher levels, as well as greater uncertainty. There may also be an increase in groundwater infiltration.
• Melting of glaciers in neighboring countries, resulting in a shift in the period of low water level on the Danube from autumn month to earlier.
• Increase in the frequency and intensity of heavy precipitation, increasing floods in inhabited areas.
• Increase in climate aridity, increase in surface temperature and heat dissipation and a decrease in precipitation will result in less drainage and water infiltration. A decrease in the annual renewable water reserves, a deterioration of the water balance of lakes and ponds, may cause the disappearance of some lakes.
• Increase in the frequency of drought, especially in the Great Plain.</td></tr>
<tr><td rowspan="2">Primary concerns</td><td>Water quantity</td><td>Water quality</td><td>Water supply and sanitation</td><td>Extreme weather events</td><td>Ecosystems</td></tr>
<tr><td></td><td></td><td></td><td>✓
(floods and droughts)</td><td></td></tr>
<tr><td>Key vulnerabilities</td><td colspan="5">• Large territories facing increased frequency of extreme flood events. Limited potential to heighten dams.
• One-third of the country's lowlands are significantly or moderately threatened by inland inundation.
• The Great Plain lacks surface water resources. In certain areas, there is no local surface water. Overexploitation of groundwater resources.
• Conflicts arising from the control of water level in large lakes (e.g. Lake Balaton).
• The quality and quantity of water flows is strongly influenced by water use and pollution in neighbouring countries.</td></tr>
</table>

Sources: Government of Hungary (2010), *Climate Change and Hungary*, Ministry of Environment and Water, Budapest; Government of Hungary (2009), *Fifth National Communication to the UNFCCC*, *http://unfccc.int/national_reports/annex_i_natcom/submitted_natcom/items/4903.php* (accessed 20 June 2012).

Key policy documents

Document	Reference to water?	Type	Year	Responsible institution
Climate Change Act		Legal act	2007	
Hungarian National Climate Change Strategy[1]	Y	National climate change strategy	2008	
National Climate Change Programme[2]		National programme		
		National adaptation plan	Under development	
Climate Change and Hungary (the VAHAVA Report)	Y	National impact assessment	2010	Ministry of Environment and Water
Local climate change strategies		Sub-national responses		Several cities[3]

1. The Climate Change Act (2007) mandates periodic revision of the National Climate Change Strategy (NCCS). The first revision is due to be completed before the end of 2013.
2. The Climate Change Act (2007) required the adoption of a National Climate Change Programme (NCCP) every two years. The first NCCP covered 2009-10.
3. Tatabánya, Hosszúhetény, Pomáz, Albertirsa, Eger, Szekszárd, Tata, XII. district of Budapest, Tata and Gyöngyös.

Policy instruments

Areas \ Policy mix	Regulatory instruments	Economic instruments	Information and other instruments
Water quantity			• Hungarian Alliance of Climate-Friendly Cities: Partnership of local governments and NGOs providing technical advice, tools, case studies and information to support cities in the realisation of climate adaptation strategies and programmes.
Water quality			
Water supply and sanitation			
Extreme weather events			
Ecosystems			

Main research programmes

- A climate change impact assessment in the Carpathian Basin was undertaken by the Regional Climate Modelling Group of the *Hungarian Meteorological Service (HMS)* together with the *Eötvös Loránd University's Department of Meteorology.*
- A vulnerability assessment for ecosystems was prepared by the *Institute of Ecology and Botany of the Hungarian Academy of Sciences.*

Principal financing mechanisms and investment programmes

- The sequence of severe floods in recent years resulted in the allocation of flood control expenditure from the central budget. Flood protection will continue to benefit from EU support in the framework of Environment and Energy Operational Programme (EEOP), for which EUR 607 million has been allocated over the period 2007-13.
- The ongoing development of a National Drought Strategy envisages the establishment of a Drought Fund.

Iceland

Climate change impacts on water systems

Observed changes and trends	• Since the 1980s, Iceland has experienced considerable warming. From 1975 to 2008, the warming rate in Iceland was 0.35 °C per decade, which is substantially greater than the global average. • Decadal variations in precipitation are significant, showing a tendency for higher amounts of precipitation during warmer periods. • Almost all of Iceland"s glaciers are receding.
Projected impacts	• Increase in temperatures at a rate of 0.16 °C to 0.28 °C per decade, resulting in a temperature rise of 1.4 °C to 2.4 °C by the end of the century. • Increase in precipitation by 5% on average. Studies suggest that precipitation increases by about 2.5% for each degree of temperature rise. • Major runoff changes resulting from the rapid retreat of glaciers. Changes in glacier runoff are one of the most important consequences of future climate changes in Iceland. The expected runoff increase may have practical implications for the design and operation of hydroelectric power plants. • Glaciers could largely disappear in the next century or two. Glaciers are a defining feature of the Icelandic landscape, covering over 10% of the island's area. • Changes in the courses of glacial rivers due to the rapid retreat of glaciers may affect roads and other communication lines. • Reduction in the number of frost days and more frequent heat waves.

Primary concerns	Water quantity	Water quality	Water supply and sanitation	Extreme weather events	Ecosystems
Key vulnerabilities					

Source: Ministry of Environment (2010), *Iceland's Fifth National Communication on Climate Change under the UNFCCC, http://unfccc.int/national_reports/annex_i_natcom/submitted_natcom/items/4903.php* (accessed 22 June 2012).

Key policy documents

Document	Reference to water?	Type	Year	Responsible institution
Iceland's Climate Change Strategy	N	National climate change strategy	2007	Ministry of Environment

Policy instruments

Areas \ Policy mix	Regulatory instruments	Economic instruments	Information and other instruments
Water quantity			
Water quality			
Water supply and sanitation			
Extreme weather events			
Ecosystems			

Main research programmes

- The Icelandic Meteorological Office (IMO) is a governmental institute responsible for producing regular and specific weather forecasts. It conducts monitoring and scientific studies of geohazards and hazard zoning in Iceland. It is involved in research within the fields of meteorology, hydrology and geosciences and has a leading role in climate change studies in Iceland. It conducts glaciological measurements and modelling with a special focus on glacio-hydrology.

Principal financing mechanisms and investment programmes

Ireland

Climate change impacts on water systems

<table>
<tr><td>Observed changes and trends</td><td colspan="5">• Warming evident in two periods, from 1910 to the mid-1940s and from 1980 to 2004. In the latter period, warming occurred at a much greater rate than the global temperature rise.
• Increases in annual and seasonal mean maximum and minimum temperatures.
• Changes to precipitation patterns are more spatially and seasonally variable than temperature changes.
• On the West Coast, significant annual increases in the number of days of extreme precipitation events (number of days where daily precipitation is greater than or equal to 10 mm).</td></tr>
<tr><td>Projected impacts</td><td colspan="5">• Increase in annual mean temperture of 1 °C to 3 °C by 2100, compared to the average of 1961-2000.
• Wetter winters in the west, drier summers in the southeast. Decrease in summer rainfall of 5% to 25% in 2021 to 2060, as compared to 1961 to 2000.
• Less snow and fewer days of snow.
• Changing patterns of precipitation will clearly impact on water service provision and may increase risk of pollution and contamination.</td></tr>
<tr><td rowspan="2">Primary concerns</td><td>Water quantity</td><td>Water quality</td><td>Water supply and sanitation</td><td>Extreme weather events</td><td>Ecosystems</td></tr>
<tr><td></td><td></td><td>✓</td><td></td><td></td></tr>
<tr><td>Key vulnerabilities</td><td colspan="5"></td></tr>
</table>

Source: Department of the Environment, Heritage and Local Government (2010), *Ireland's Fifth National Communication under the UNFCCC*, *http://unfccc.int/national_reports/annex_i_natcom/submitted_natcom/items/4903.php* (accessed 22 June 2012).

Key policy documents

Document	Reference to water?	Type	Year	Responsible institution
National Climate Change Strategy	Y	National climate change strategy	2007-12	Department of the Environment, Community and Local Government (DECLG)
National Climate Change Adaptation Framework	Y	National adaptation framework	2012	DECLG
A Summary of the State of Knowledge on Climate Change Impacts for Ireland		National impact assessment	2009	Environmental Protection Agency (EPA)
National Adaptive Capacity Assessment	Y	National adaptive capacity assessment	2012	DECLG, EPA
Climate Change Strategy for Dublin City 2008-12	Y	Sub-national responses	2008-12	The Environment and Engineering Strategic Policy Committee in association with City of Dublin Energy Management Agency

Policy instruments

Areas \ Policy mix	Regulatory instruments	Economic instruments	Information and other instruments
Water quantity			• National Catchment-based Flood Risk Assessment and Management Programme: Involves consideration of potential future climate change scenarios, as well as maps of existing conditions. This programme is now well underway following completion of a set of pilot projects, *www.cfram.ie*. • Flood information websites: As part of the strategy to manage flood risk, the website aims to raise awareness of flood risk and promote preparedness, effective emergency response planning and better flood forecasting and warning, *www.flooding.ie*. • The National Climate Change Adaptation Framework provides a clear mandate for the preparation of an adaptation plan for the water sector.
Water quality			
Water supply and sanitation			
Extreme weather events	• SI No. 122 of 2010, European Communities (Assessment and Management of Flood Risks) Regulations 2010: Sets requirements in relation to climate change.		
Ecosystems			

Main research programmes

- Climate Change Research Programme of the Environmental Protection Agency: Aims to advance the understanding of and to support action to address climate change. To support adaptation and risk management. It aims to provide information on future climate impacts and vulnerability, *www.epa.ie/researchandeducation/research/ourresearchprogramme/climatechange/#d.en.33770*.
- Project on "Co-ordination, Communication and Adaptation for Climate Change in Ireland": Aims to identify an effective integrated approach to vulnerability assessment and adaptation in key sectors, including water.

Principal financing mechanisms and investment programmes

Israel

Climate change impacts on water systems

Observed changes and trends	• Annual rainfall has decreased by 9% on average since 1993. • An increase in the frequency and duration of extreme weather events has been observed in recent years, including years that were either exceedingly wet or exceedingly dry. • Nearly seven consecutive years of drought were experiences from 2003 to 2010.
Projected impacts	• Increase in average annual temperatures of between 0.3 °C to 0.5 °C per decade. • Decrease in precipitation by 10% by 2020 and by 20% by 2050. • Reduction of at least 25% in water availability from 2070 to 2099, as compared to 1961 to 1990. • Reduced flows to Lake Kinneret. • Reduced recharge of groundwater aquifers and negative impacts on freshwater ecosystems. • Further increases in the number and frequency of extreme weather events (e.g. drought years, floods, heat waves), which will result in damage to property and ecosystems. • Increase in the desertification of the southern part of Israel. • Sea level rise in the Mediterranean by 0.5 meters in 2050 and one meter by 2100. • Changes in the salinity level of the Sea of Galilee.

Primary concerns	Water quantity	Water quality	Water supply and sanitation	Extreme weather events	Ecosystems
	✓	✓		✓	

Key vulnerabilities	• Israel is characterised by arid and semi-arid climatic conditions and water scarcity is already a major concern. • Israel already consumes more water than its natural supply (essentially provided by rainfall), with an annual deficit of 300 million m^3/year as of 2011. • No water from the Jordan River now reaches the Dead Sea, which has lost one-third of its surface area since 1930.

Sources: Ministry of Environmental Protection (2010), *Israel's Second National Communication on Climate Change*, *http://unfccc.int/national_reports/annex_i_natcom/submitted_natcom/items/4903.php* (accessed 21 June 2012); Ministry of Environmental Protection (2009), *Coping with Climate Change in Israel*, *http://old.sviva.gov.il/bin/en.jsp?enPage=e_BlankPage&enDisplay=view&enDispWhat=Object&enDispWho=Articals^l7003& enZone=knowledge_center* (accessed 11 November 2012); OECD (2011), *Environmental Performance Review: Israel 2011*, OECD Publishing, Paris, *http://dx.doi.org/10.1787/9789264117563-en*.

Key policy documents

Document	Reference to water?	Type	Year	Responsible institution
		National adaptation strategy	Under development	
		National adaptation plan	Under development	

Policy instruments

Areas \ Policy mix	Regulatory instruments	Economic instruments	Information and other instruments
Water quantity			• Israel Climate Change Knowledge Centre: Established by the Ministry of Environmental Protection in 2011, it aims to gather and co-ordinate scientific knowledge available on climate change. Information gathered will be incorporated in policy documents and will contribute to the formulation of a national plan on adaptation to climate change, *http://old.sviva.gov.il/bin/en.jsp?enPage=e_BlankPage&enDisplay=view&enDispWhat=Zone&enDispWho=knowledge_center&enZone=knowledge_center.*
Water quality			
Water supply and sanitation		• Water pricing: To promote more efficient water use, tariffs for the domestic sector were raised by 40% in 2010.	
Extreme weather events			
Ecosystems			

Main research programmes

- Israel Climate Change Knowledge Centre: (see description above), *http://old.sviva.gov.il/bin/en.jsp?enPage=e_BlankPage&enDisplay=view&enDispWhat=Zone&enDispWho=knowledge_center&enZone=knowledge_center.*

Principal financing mechanisms and investment programmes

Highlights and innovative initiatives

- **Reuse of treated domestic wastewater** (effluent): To ensure high-quality effluent for irrigated crops, soil aquifer treatment (SAT) has been used at the Dan wastewater treatment and reclamation plant in Tel Aviv (130 million m^3/year or about 25% of total wastewater treated). Effluent is discharged to sand infiltration fields, where physical, biological and chemical processes treat it before reaching the aquifer.
- **Large-scale desalination of seawater, maximising energy efficiency:** Three large-scale desalination facilities currently provide 320 million m^3 of potable water to all sectors. By 2015, 2025 and 2050, respectively, new desalination facilities along the southern end of the country's Mediterranean coast are expected to cover 62.5%, 70% and 100% of domestic water demand. Energy efficiency is maximised through the bidding process for the construction of these plants. Energy efficiency is promoted by giving preference to natural gas (rather than use of coal) and to efficient technological energy recovery systems. Israel's desalination water production is therefore among the most energy-efficient (3.5 kWh/m^3) and cost-efficient (USD 0.54/m^3) in the world. While large-scale desalination significantly increases water availability, it has potentially adverse environmental impacts, in particular in the form of greater energy consumption and thus, increases in greenhouse gas emissions.

Italy

Climate change impacts on water systems

Observed changes and trends	• 2008 was considerably warmer than average, compared with the period 1961-90. • Increase in rainfall by 20% in 2008 in northern and central Italy, thus in it was the third wettest year of the period 1961 to 2008 in central Italy. At the same time, there was decrease by 7% in 2008 in the South and in the islands, as compared to the average rainfall in the period 1961 to 1990.
Projected impacts	• Deterioration of the existing conditions of high stress on water resources leading to reduced water availability and quality especially during summer in southern regions and small islands. Water stress may increase by 25% in this century with a growing demand for irrigation water. • Increase in water demand, lack of adequate management practices, aggravated by further decreases in mean precipitation could result in challenges to ensure safe water supply in several regions (e.g. Puglia, Basilicata, Sicilia and Sardegna). • Reduced availability of water resources, impacting on water supply for household and irrigation and for hydropower generation in the Po river valley. Reduction of water availability in the North and in the Centre of Italy. • Water quality depletion due to saltwater intrusion into coastal freshwater aquifers, loss of wetlands, and temperature increase with impacts on lake water. • Alterations of the hydro-geological regime putting more than 5% of the national territory at risk of floods and landslide due to severe precipitation. • Shifts in the Alpine water regime due to changes in precipitation, snow-cover patterns and glacier storage. By further modifying run-off regimes these impacts will lead to more droughts in summer, floods and landslides in winter and higher inter-annual variability. In the Italian central Alps, rivers could experience an increased winter run-off by 90% and a decreased summer run-off by 45%. • Glacial lake outburst flooding due to glacier melting in the Alpine area. • Increase of flash-flood events and debris flows due to the increase in extreme event in mountainous areas. • Increase of flood events in the southern regions. • Increase in the frequency of droughts and increase in soil dryness in the areas of the plains. • Navigation of lakes and rivers impaired by a reduction of precipitation and water levels. • Loss of Alpine biodiversity, such as forests and glaciers ecosystems, also important for tourism.

Primary concerns	Water quantity	Water quality	Water supply and sanitation	Extreme weather events	Ecosystems
	✓			✓ (droughts in southern regions; hydro-geological risks in the Alpine area)	✓ (loss of biodiversity)

Key vulnerabilities	• Italian Alpine regions are particularly vulnerable. • The most vulnerable area for floods and landslides is the Po River Basin. • Southern regions are already suffering from widespread water stress and local soil degradation.

Source: Ministry for the Environment, Land and Sea (2009) *Fifth National Communication under the UNFCCC*, *http://unfccc.int/national_reports/annex_i_natcom/submitted_natcom/items/4903.php* (accessed 21 June 2012).

Key policy documents

Document	Reference to water?	Type	Year	Responsible institution
National Adaptation Strategy to Climate Change Impacts		National adaptation strategy	Under development	Ministry for the Environment, Land and Sea (IMELS)
National Plan of Agrarian Biodiversity[1]	Y	National biodiversity plan	2012	Ministry of agricultural food and forestry policies

1. Biodiversity is defined as the "variability of living beings in terrestrial and water ecosystems", therefore the actions about biodiversity involve the environmental status of water ecosystems.

Policy instruments

Areas \ Policy mix	Regulatory instruments	Economic instruments	Information and other instruments
Water quantity			• Rural development measures aimed at saving water, best practice water management and the defence of biodiversity. • Integrated national and regional early warning system for hydro-geological and hydraulic risks for the purpose of civil protection.[1] • National Action Plan to combat drought and desertification[3] includes measures to address water and groundwater protection and water efficiency through planning instruments and water protection plans. • Pilot projects for adaptation: Under the guidance of the National Committee to combat drought and desertification, lead to the development of 6 Local Action Plans to combat drought and desertification. • National Plan for Biodiversity: Includes measures to address biodiversity protection, including water ecosystems protection.
Water quality			
Water supply and sanitation			
Extreme weather events	• Regulations for water emergencies: To address water crises, providing technical and financial support for emergency measures. • Law 267/1998[2] ("*Legge Sarno*"): Requires water basin authorities to detect risk areas, set prevention plans and establish regulations to avoid additional risk. • Directive 2007/60/CE implemented in Italy with Decree 152/2006: Aims to establish a framework for the assessment and management of flood risks, aiming at the reduction of the adverse consequences for human health, the environment, cultural heritage and economic activity associated with floods in the Community. The Legislative Decree No. 49/2010 requires flood risk management plans to be established and approved by June 2015 (under implementation).		
Ecosystems			

1. Established by the Prime Minister Directive, 27 February 2004.
2. Approved by the Inter-ministerial Economic Planning Committee (CIPE) with Deliberation No. 229, 21 December 1999.
3. Law 267/1998 establishes the legal basis for the identification and funding of urgent preventive measures.

Main research programmes

- Euro-Mediterranean Centre on Climate Change: National Research Centre on climate science and policy undertakes integrated, multi-disciplinary and frontier research for understanding, controlling and adapting to climate change, *www.cmcc.it*.
- Agroscenari project: Through an integrated analysis of Italian agricultural systems in possible future climate scenarios, Agroscenari aims at developing cognitive and decision making tools for supporting agricultural activities. The research will enable adaptation to climate change, according to environmental and socio-economic sustainability criteria, and considering the increasing economic value of water resources, *www.agroscenari.it*.

Principal financing mechanisms and investment programmes

- Reducing the risk of floods and landslides: The total cost of reducing the risk of floods and landslides in Italy is estimated at EUR 42 billion (of which only EUR 1.15 billion were budgeted for in 2006). However, this estimate does not take into account the higher risks deriving from climate change scenarios, for which no assessment currently exists. The Inter-Ministerial Committee for Economic Planning has committed EUR 1 billion in 2009 to the Ministry for Environment, Land and Sea for extraordinary operations concerning hydro-geological instability.
- Rural development programmes: Measures aimed at saving water, good water management and addressing hydro-geological instability amounts cost EUR 54 million over the period 2007-11. These measures were co-funded by the European Regional Development Fund (ERDF).
- National Solidarity Fund for natural disasters in agriculture. This fund partially compensates farmers for damage due to extreme events, such as floods and droughts. The financial allocation to the Fund varies yearly in relation to the availability of funding from the annual State budget.

Japan

Climate change impacts on water systems

Observed changes and trends

- Restrictions on water supply (including stoppages on tap water supply) due to unprecedented low rainfall.
- Reduction in the amount of water in dams, which has potential to supply water in a stable manner.
- Abnormal occurrence of blue-green algae in lakes, with impacts on water supply and freshwater ecosystems.
- Significant increase of flooding, sediment-related disasters and other water-related disasters due to frequent daily rainfall amounts from 100 mm up to 200 mm over the past 100 years.
- Frequent occurrences of localised heavy rainfall exceeding 80 mm per hour in recent years, leading to large-scale flooding and sediment-related disasters in various areas throughout Japan on a yearly basis.
- Increased use of groundwater during drought leading to the occurrence of ground sinking.
- Shrinking freshwater areas suitable for cold-water fish.

Projected impacts

- Increase in annual mean temperature across Japan of approximately 2 °C to 4 °C from the end of 20th century to the end of the 21st century.
- Shift in the number of very warm and very cold days. For example, a reduction in the number of days with minimum temperature less than 0 °C and an increase in the number of nights with minimum temperature of 25 °C or higher.
- Considerable fluctuation in the amount of precipitation with an overall increase in the amount of annual rainfall.
- Increase in the frequency of heavy rainfall with daily precipitation of 100 mm or more. Also, an increase in the number of days with no precipitation.
- Decrease in the amount of snowfall from Tohoku to the Sanin region, mainly on the Japan Sea side.
- Groundwater salination due to sea-level rise.
- Increased water temperature of rivers, lakes, dammed lakes, and groundwater. Higher probability of increase in blue-green algae outbreaks.
- Notable decrease in the flood safety control level[1] (which indicates the degree of safety of rivers in flood-control plans) if precipitation levels rise in the future. This will increase the danger of flooding and over spilling in watershed areas.
- Increased risk of drought.
- Changes in species distribution and entry of new exotic species in freshwater areas.

Primary concerns	Water quantity	Water quality	Water supply and sanitation	Extreme weather events	Ecosystems
				✓	

Key vulnerabilities

1. For instance, it is predicted that the frequency of the flood safety control level targeted in current plans for rivers will drop from once every 100 years to once every 30 years, thereby increasing the frequency of flooding three-fold.

Sources: Committee on Climate Change Impacts and Adaptation Research (2008), *Wise Adaptation to Climate Change*, Ministry of the Environment, Tokyo; Expert Committee on "Synthesis Report on Observations, Projections, and Impact Assessments" (2009), *Climate Change and its Impacts in Japan*, Ministry of Education, Culture, Sports, Science and Technology, Japan Meteorological Agency, and Ministry of Environment, Tokyo; Government of Japan (2009), *Japan's Fifth National Communication under the UNFCCC, http://unfccc.int/national_reports/annex_i_natcom/submitted_natcom/items/4903.php* (access 21 June 2012); Japan Meteorological Agency (2010), *Climate Change Monitoring Report, www.jma.go.jp/jma/en/NMHS/indexe_ccmr.html* (accessed 28 June 2012).

Key policy documents

Document	Reference to water?	Type	Year	Responsible institution
Synthesis Report on Observations, Projections and Impact Assessments of Climate Change: Climate Change and Its Impacts in Japan	Y	National impact assessment	2009	Ministry of Education, Culture, Sports, Science and Technology, Japan Meteorological Agency, Ministry of Environment
Wise Adaptation to Climate Change		Extensive review document	2008	Ministry of the Environment
Approaches to Climate Change Adaptation		Report	2010	Ministry of the Environment

Policy instruments

<table>
<tr><th>Areas \ Policy mix</th><th>Regulatory instruments</th><th>Economic instruments</th><th>Information and other instruments</th></tr>
<tr><td>Water quantity</td><td></td><td></td><td rowspan="5">• Watershed-based water supply development master plans.
• Guidance document "Approaches to Climate Change Adaptation": Published in 2010, aims to help policy makers design and evaluate adaptation responses to climate risks.
• Promotion of rainwater and reclaimed water use will be promoted in order to respond to the risk of drought, which will increase due to climate change.
• Regional disaster response and disaster-prevention plans: Including the provision of hazard maps, enhancing river information for flood warnings.
• Weather and land information for warnings of sediment-related disasters.</td></tr>
<tr><td>Water quality</td><td></td><td></td></tr>
<tr><td>Water supply and sanitation</td><td></td><td></td></tr>
<tr><td>Extreme weather events</td><td>• Regulations and guidance on land use in disaster-prone areas, as well as unified flood control measures.</td><td></td></tr>
<tr><td>Ecosystems</td><td></td><td></td></tr>
</table>

Main research programmes

Principal financing mechanisms and investment programmes

Korea

Climate change impacts on water systems

Observed changes and trends	• Average temperature has been rising gradually over the past 90 years. The temperature rise has been the highest in the 1990s. • Frequency of extremely high temperatures in the summer has slightly increased. • Precipitation shows an increasing trend. The number of rainy days decreases while rainfall intensity tends to increase. Days with heavy rainfall of over 50 mm have increased by approximately 22% to 25%. Rainfall intensity has increased by 18% in the southern region over the past 20 years. • Precipitation run-off has surged due to increasing impervious areas caused by urbanisation. • The amount of evapotranspiration has a decreasing trend across the country, in spite of increasing precipitation and temperature. • Extreme climate events such as drought, high temperature and torrential rainfall have become more frequent (e.g. in 2012, monthly rainfall from May to June was 30% below normal; there was scorching heat for 17 consecutive days from 24 July to 9 August; and 444 mm of torrential rainfall in Gunsan on 13 August). • Increase in the magnitude and frequency of floods. In 1996, 1998 and 1999, there were outbreaks of intense and extended rainfall over wide areas in the northern region of Gyeonggi province, causing floods. The damage was aggravated due to infrastructure (roads, bridges and structures) that were constructed without consideration of drainage to counter flood damage. • Emergency water rationing was implemented for 400 000 people for 10 years due to the occurrence of winter drought.
Projected impacts	• Increase of annual precipitation in 6 major cities by 19% over the next 100 years. • Two-fold increase in rainfall intensity, as compared to the 1970s. • Increased likelihood of algae outbreak in public waters due to climate change impacts such as increasing number of scorching hot days and changes in rainfall patterns. • Increased likelihood of water quality degradation due to non-point source resulting from increasing amounts and severity of precipitation. • Increase in the variability of discharge. Drought in Korea is caused by massive seasonal and yearly variations in discharge. Thus, even an increase in rainfall may not contribute to the relief of water shortage if the shift in variability in discharge increases. It may, instead, worsen the shortage. • Increase in the frequency of longer droughts and higher temperatures generate algae blooms in water systems that make it more difficult to control the quality of tap water. • Increase in risk of flood. As climate change causes an increase in rainfall, the concentration of rainfall in the summertime will increase the frequency and severity of floods. Increase in flooding in urban areas, due to infrastructure lacking adequate drainage to respond to increasing severity of intense rainfall. Increase in the frequency of urban flooding also resulted from aggravated water circulation, depletion of groundwater and drying-up of rivers. • Changes in the distribution of aquatic ecosystems caused by changes in water temperature and volume as a result of climate change.

Primary concerns	Water quantity	Water quality	Water supply and sanitation	Extreme weather events	Ecosystems
	✓ (urban flooding)	✓ (algae outbreak, non-point source pollution)	✓ (threat to safety of tap water due to algae outbreak)	✓ (threat to safe urban life due to torrential rain)	✓

Key vulnerabilities	• Rise in inefficiency and cost of water purification process to reduce odour resulted from algae. • Korea's rivers and lakes are hyper eutrophic, so when the certain climate conditions are met (e.g. temperature, amount of sunshine), the algae blooms may occur in public waters systems. • Facing limits in managing water quality and volume due to changes in precipitation patterns and rising water temperature. • Many infrastructures (roads, bridges and structures) have been constructed without consideration of extreme floods due to climate change, which aggravates flood severity. • Ecosystems are especially vulnerable in the subtropical zone and mountainous areas. • There are institutional frameworks (e.g. legal acts) for recreational water value and the efficient use of water, but a similar institutional framework for infrastructures is still in its infancy.

Source: Ministry of Environment (2012), *Korea's Third National Communication under the United Nations Framework Convention on Climate Change: Low Carbon, Green Growth*, *http://unfccc.int/national_reports/non-annex_i_natcom/items/2979.php* (accessed 20 June 2012).

Key policy documents

Document	Reference to water?	Type	Year	Responsible institution
Framework Act on Low Carbon, Green Growth (2009-50)		Legal act	2010	Presidential Committee on Green Growth
National Climate Change Adaptation Master Plan[1] 2011-15	Y	National adaptation strategy	2010	Ministry of Environment, Ministry of Land, Transport and Maritime Affairs and other related 11 Ministries.
Vulnerability assessment for each of 13 sectors		Vulnerability assessment	Under development	
Local Government Action Plan		Sub-national responses	2012	
Master plan for Responding to Climate Change and Improving Disaster Management	Y	National adaptation plan for disaster management	2011	Office of Prime Minister, Ministry of Environment and other related 9 Ministries

1. Based on the Master Plan, several specific plans and schemes are established by relevant ministries.

Policy instruments

Areas \ Policy mix	Regulatory instruments	Economic instruments	Information and other instruments
Water quantity	• Requirement to construct rainwater-equipped facilities when building a new sports complex, gym and/or government office building, or extend an existing building by more than 1 000 m^2. • Approval of reuse of treated water for public water treatment facilities that treat more than 5 000 m^3 of water per day (approved since 28 September 2008).		• Early warning systems: Efforts to establish a systematic and efficient structure that will provide early warning to the central government and local authorities to enhance the efficiency of water resource management and minimise damage from disasters. • Korea Adaptation Centre for Climate Change: Supports central and local governments to develop adaptation measures to climate change and provide guidelines for policy issues associated with climate change adaptation. The Centre also operates the Climate Adaptation Information Delivery System and a website for disseminating information on climate change adaptation, *http://kaccc.kei.re.kr/english/eng_index.do*. • Guidelines for using algae removal facility and spraying algae-removing substances.
Water quality	• Registration to construct facilities to reduce non-point source pollution. • Designation of areas of restricted livestock raising, strengthening of nitrogen and phosphorus standards in effluence, adoption of e-transfer system to monitor transport of wastewater from discharge to collection to delivery to treatment. • Water Quality and Ecosystem Conservation Act: Includes climate adaptation measures for water quality and ecosystem conservation plan for significant areas. • Second Master Plan on the management of non-point source pollution. • Master Plan on advancing animal waste Management.		
Water supply and sanitation	• Registration of construction of treated water supply (target) for new building projects, such as accommodations with over 60 000 m^2 of total floor area, factories that generate over 1 500 m^3 of waste water a day, the development business of housing sites, industrial complex, and tourist facilities, etc., that are run by the State or a public corporation. • Act on the Promotion and Support for Water Re-use.		

Policy instruments (cont.)

Areas \ Policy mix	Regulatory instruments	Economic instruments	Information and other instruments
Extreme weather events		• Small stream restoration project. • Disaster prone area enhancement project.	
Ecosystems			

Main research programmes

- Evaluation of climate change adaptation policies and establishment of climate change adaptation policy inventory: Aims to assess existing policies to determine which policies have considered climate change adaptation. It also establishes a policy inventory.
- Research on incorporating climate change adaptation considerations in Environmental Impact Assessment: aims to assess the impacts of climate change at the project level, to identify climate vulnerability, and ultimately to suggest the guidelines for incorporating climate change adaptation into the EIA process.
- Research to study climate change adaptation policy to improve the legal framework: Examines the policy and legal system to improve consideration of the climatic conditions of Korea.
- Targeting system for managing water demand: Mayors and governors of local governments are encouraged to establish a master plan on water demand management every 5 years and get approval from the Minister of Environment in order to enhance the efficiency of waterworks and strengthen demand management for tap water.
- Comprehensive research on algae outbreaks, eco-friendly algae reduction technology and toxic algae control.
- Enhanced monitoring of aquatic ecosystem for climate adaptation.
- Research on building aquatic ecosystem clinic program (2012-17): By identifying causes of climate change impacts on aquatic ecosystem and finding restoration solutions, the research aims to improve the effectiveness of monitoring system and restore the health of aquatic ecosystems.
- Research group for non-point source pollution focuses on the development of technology to reduce run-off and non-point source pollution.

Principal financing mechanisms and investment programmes

- Four Rivers Restoration Project (4RR): Aimed to secure enough water to respond to future water scarcity and severe drought due to climate change (target: water supply of 1.3 billion m^3); to improve flood control capacity (target: 920 million m^3); to improve water quality (target: BOD 3 mg/l); to restore ecological rivers, create wetlands and readjust farmland to rehabilitate the ecosystems (223 restoration projects planned); and develop rivers banks to provide space for leisure and develop regions around rivers. Korea invested USD 17.6 billion to implement the 4RR, which was completed in 2011.
- Examples of several State funding mechanisms include:
  - State funding for construction of a facility that uses rainwater or treated reused water.
  - State funding for using a "Low Impact Development".
  - State funding for constructing a facility that reduces non-point source pollution; increased governmental subsidy (from 50% to70%) for designating a management area for non-point source pollution.
  - State funding for constructing a facility to treat public animal waste treatment.
  - State funding for water supply maintenance preparing for torrential rain.

Highlights and innovative initiatives

- **"Act on Water Reuse"** (2011): Facilitates the efficient use of secured water resources. The Act responds to challenges arising from the expected increase in demand for water due to rising living standard and economic activities, while the supply of water resources may face further pressures from climate change. The Act establishes the institutional framework to reduce dependency on existing freshwater resources and to reuse rainwater, waste water and treated water. As such, it can contribute to addressing water shortage caused by climate change and encouraging efficient use of water resources.
- **Initiatives to address urban flooding** (2011-12): Strengthens the standard for sewer pipe facilities to tolerate heavy rainfall events with a 30 year return period (previous return period of 10 years), designates flooded areas as the main target for management, makes it mandatory to construct rainwater reusing facility for buildings over certain size.
- Introduction of scientific index for status evaluation of eutrophication of rivers and lakes (chl-a, T-P, COD, etc.); strengthening of monitoring of algae drivers and R&D.
- Introduction of performance certification of facilities for the reduction of non-point source pollution.
- Establishment of a clinic program for aquatic ecosystem health (2012-17).

Luxembourg

Climate change impacts on water systems

Observed changes and trends	• Increase in annual mean temperatures over the last decades. The annual mean temperature for the capital that was typical for the second half of the 20th century (around 9 °C) is now regularly exceeded. Since 2000, the annual mean temperature is between 9.3 °C (2001) and 11.3 °C (2007). This rise is mainly driven by higher temperatures in winter. • Seasonal distribution of precipitation has shown substantial variability over the past 130 years. • Increased frequency of floods due to major redistributions of winter rainfall has caused an increase in maximum daily runoff during winter. • Increase in the frequency and length of dry periods.
Projected impacts	• Increase in temperature from 1 °C to 2.2 °C by 2050, as compared to the period 1960-90. Annual mean temperature is expected to reach up to 11.6 °C for the period 2071 to 2100. • Relatively stable annual totals of precipitation up to 2100 suggested by preliminary studies. However, a substantial redistribution of seasonal precipitation totals can be expected in the second half of the century, with a decrease in summer rainfall and an increase in winter precipitation. In winter, increase in rainfall from 0 to 25% by 2050 with increasing discharges. In summer, decrease in rainfall of 5 to 25% with reduced run-off by 2050. • Deterioration of water quality due to intensification of rainfalls (increasing erosions, rapid infiltration towards groundwater) and shift in the main recharge period of groundwater. • Changes in the water cycle could increase public health risks related to water quality and water scarcity. • Increase in the risk of river flood in winter and an increase in the risk of droughts.

Primary concerns	Water quantity	Water quality	Water supply and sanitation	Extreme weather events	Ecosystems
	✓	✓	✓	✓ (floods)	

Key vulnerabilities	• Drinking water supply due to an expected increase in drinking water consumption as a result of demographic changes.[1] • Groundwater recharge due to a shift of the main recharge period. • Changes in the hydrological regime of rivers. • Water quality deterioration due to an intensification of rainfall, increasing erosion, rapid infiltration of groundwater, etc. • Flood and low water management. • Navigation.

1. Possible restrictive water use measures to be put in place during warm periods in order to manage water consumption peaks. Drinking water supply also has to deal with 150 000 daily commuters from bordering countries.

Sources: Department of Environment (2010), *Fifth National Communication of Luxembourg, http://unfccc.int/national_reports/annex_i_natcom/submitted_natcom/items/4903.php* (accessed 20 June 2012); Görgen, K. et al. (2010), *Assessment of Climate Change Impacts on Discharge in the Rhine River Basin: Results of the RheinBlick2050 Project*, CHR report, I-23, 229 pp., Lelystad, ISBN 978-90-70980-35-1; International Commission for the Protection of the Rhine (2011), *Study of Scenarios for the Discharge Regime of the Rhine, www.iksr.org/fileadmin/user_upload/Dokumente_en/Reports/188_e.pdf* (accessed 12 August 2012).

Key policy documents

Document	Reference to water?	Type	Year	Responsible institution
National Adaptation Plan[1]	Y	National adaptation plan	2011	

1. Adopted in June 2011 by Luxemboug's Council of Ministers, the plan prioritises biodiversity, water, agriculture and forestry.

Policy instruments

Areas / Policy mix	Regulatory instruments	Economic instruments	Information and other instruments
Water quantity	• Licensing and metering of surface and groundwater abstractions. • Prohibition of certain water uses (e.g. irrigation) in order to guarantee water supply during critical periods (e.g. "*Phase orange*", "*Phase rouge*").		• Efforts to reduce the inefficient water use and promote water saving infrastructures and rainwater catchment systems. • Campaigns to raise awareness about saving drinking water. • Protection against erosion, development of sustainable soil use. • Preliminary flood risk assessment has been undertaken along with the preparation of flood hazard maps and flood risk maps. The establishment of a flood risk management plan is planned for 2015. • Early warning systems: Luxembourg has a dense national network of meteorological stations and hydrological stations and initiatives for the public dissemination of information (competent authorities and the public are informed and warned about floods ±48 hours before the flood event.[1] • Additional natural retention volume to promote prevention of flash floods. • Efforts to avoid additional stress on water resources that may result from new surface and groundwater water pollution sources due to new energy strategies (geothermal drillings, CO_2 storage, shell gas exploitations, bio fuel).
Water quality			
Water supply and sanitation	• Mandatory measures to limit drinking water peak consumption during warm summer periods (e.g. "*Phase orange*", "*Phase rouge*").		
Extreme weather events			
Ecosystems			

1. The Flood warning service is operated in co-operation with the Water Management Agency and the Water navigation agency, *www.inondations.lu* (in French).

Main research programmes

- The Luxembourg Government supports research institutions to carry out climate change research related to vulnerability, mitigation and emission projections.
- Luxembourg is currently participating in the regional Flow MS (Flood and Low Water Management Moselle-Sarre) project to assess the consequences of climate change for floods and low water flow in the Moselle and Saar catchments and to develop adjustment strategies.[1]
- Adapting monitoring strategies in order to establish a long time series of surface and groundwater data to separate natural variability, human influences and climate change effects.

1. This project is supported by the Interreg IV-A program "Greater Region" (the Interreg Programme in the European Union: European Territorial Cooperation, commonly known as INTERREG, represents one of the pillars of this policy for the period 2007-13. It is financed by the European Regional Development Fund).

Principal financing mechanisms and investment programmes

Mexico

Climate change impacts on water systems

Observed changes and trends	• Annual mean temperature has risen on average 0.6 °C over the last four decades. • Intense rainfall in 2010 marked the second rainiest year on record. • Intense drought in 2011 to 2012 was the worst over the last 70 years. July 2009 was the second driest July with an average monthly rainfall of 99.1 mm in the period between 1941 and 2009.
Projected impacts	• Increase in mean temperature, with the highest warming in the North and Northwest. • Most projections show a decline in average annual rainfall, although the percentage reduction varies significantly between models. The reduction in rainfall impacts on runoff in rivers, water stored in dams, and aquifer recharge. • Diminished water quality as a result of a rise in temperatures. • Salt water intrusion in both surface and groundwater. • Increased severity of drought, especially in Central, Jalisco and Chiapas region. • More frequent and intense *El Niño* and *La Nina* phenomena. • Sea level rise.

Primary concerns	Water quantity	Water quality	Water supply and sanitation	Extreme weather events	Ecosystems
	✓		✓ (major cities, competition among uses)	✓ (droughts and floods)	✓ (competition among uses)

Key vulnerabilities	• Overexploitation of aquifers and over allocation of surface water in the context of significant demand from industrial and agricultural production, and growing urban areas.

Sources: Galindo, L. (2009), *The Economics of Climate Change in Mexico: Synopsis*, Ministry of the Environment and Natural Resources (SEMARNAT), Mexico City; Ministry of Environment and Natural Resources (2009), *Fourth National Communication of Mexico to the UNFCCC*, *http://unfccc.int/national_reports/non-annex_i_natcom/items/2979.php* (accessed 20 June 2012); National Water Commission (CONAGUA) (2008), *National Water Program 2007-2012*, *www.conagua.gob.mx/home.aspx* (accessed 12 August 2012).

Key policy documents

Document	Reference to water?	Type	Year	Responsible institution
National Strategy on Climate Change (ENACC)	Y	National climate change strategy	2007	Inter-ministerial Commission on Climate Change (CICC)
National Development Plan (PND)		National development plan	2007-12	Federal Institutions
Special Programme on Climate Change (PECC)	Y	Programme to implement the National Climate Change Strategy	2009-12	Federal Institutions
National Water Program (PNH)	Y	National water program	2007-12	National Water Commission (CONAGUA)
2030 Water Agenda	Y	Strategy document	2011	Water agencies at federal, state and municipal levels, as well as Society at all sectors
National Strategy for Water Resources and Climate Change (NSWR&CC)	Y	Strategy document	Under review	National Water Commission (CONAGUA)
The Economics of Climate Change in Mexico[1]		National assessment on climate change impacts and costs	2009	At the request of Ministers of Finance and Ministers of Environment and Natural Resources (SEMARNAT)

Key policy documents (cont.)

Document	Reference to water?	Type	Year	Responsible institution
Prioritisation process for "Joint Grant Contribution Program for Drinking Water and Wastewater Infrastructure Projects for Communities in the US-Mexico Border Area"	Y	Transboundary responses	2011-12	Border Environment Cooperation Commission (BECC)

1. Study co-ordinated by Dr. Luis Miguel Galindo Paliza of the Faculty of Economics of the National Autonomous University of Mexico at the request of the Ministers of Finance and for the Environment and Natural Resources.

Policy instruments

Areas \ Policy mix	Regulatory instruments	Economic instruments	Information and other instruments
Water quantity	• Lerma Chapala Surface Water Allocation Agreement:[1] Reformed existing allocation arrangements to restore sustainable levels of abstraction and adequate levels for environmental use.		• Capacity building for CONAGUA staff on climate change adaptation includes training course on climate change adaptation, technical handbooks, and operation manuals with climate change information. • Water Reserves for the Environment: Technical studies in pilot basins to be launched.
Water quality			
Water supply and sanitation			
Extreme weather events	• Drought guidelines require Basin Councils to prepare and implement actions before, during and after any drought.	• Promoting insurance as an aid to reducing vulnerability is identified as one of the priorities for adaptation.	
Ecosystems			

1. Ratified in January 2005 by the Governors of the states of Mexico, Queretaro, Guanajuato, Michoacan and Jalisco and water stakeholders.

Main research programmes

- Research programme on the impacts of climate change on thirteen hydrological administrative regions.

Principal financing mechanisms and investment programmes

- Fund for Disaster Prevention (FOPREDEN) to supports action for integrated risk management.
- Adaptation Contingency Fund: The 2030 Water Agenda proposes to establish an Adaptation Contingency Fund that would improve Mexico's capacity to effectively replace or significantly modify water supply systems and flood systems. CONAGUA is still analysing alternatives for implementing the Fund. The recently adopted General Law for Climate Change specifies the need to create a fund for projects, studies, actions. In addition, the Mexican Institute of Water Technology (IMTA) was accredited as the National Implementing Entity of the UNFCCC's Adaptation Fund.
- Joint investment programme (Mexico-USA) on the Bravo River: Under the "Joint investment programme (Mexico-USA) on the Rio Grande" each project goes through a process involving four main institutions: EPA, BECC, NADB and CONAGUA. Every two years a new process starts with a call for registration of projects to address needs related to water supply, sewerage or sanitation. Projects are then selected through a prioritisation process and certified based on environmental studies (among others) on both sides of the border.
- Colorado river joint co-operation projects to face water scarcity due to climate change: Includes investments in a bi-national desalination plant (cost: USD 140 million), expansion of the Colorado-Tijuana aqueduct from 4.0 to 5.3 m^3/s (cost: USD 150 million), aquifer recharge at Mexicali valley, modernisation and conservation of Irrigation District Colorado river (conserving 767 hm^3/year at a cost of USD 936 million) and Irrigation District 014 (conserving 89 hm^3/year at a cost of USD 85.7 million), wastewater reuse "Las Arenitas" (Mexicali II System) (cost: USD 18 million) which is now in place for irrigation. These projects are at various stages of implementation.

Highlights and innovative initiatives

- **Lerma Chapala Surface Water Allocation Agreement:** Since the implementation of the Agreement in 2005, Lake Chapala levels have shown remarkable recovery, *www.conagua.gob.mx/LermaWeb.*
- **Water Reserves for the Environment:** Technical studies in pilot basins to be launched. Funding secured in 2012. A visualisation tool is available at: *http://sigagis.conagua.gob.mx/RESERVAS%20POTENCIALES%20DE%20AGUA%20PARA%20EL%20MEDIO%20AMBIENTE.*

The Netherlands

Climate change impacts on water systems

Observed changes and trends	• Average temperature is rising faster than the global average. • Increase in annual precipitation. • Climate change is particularly felt in the risk of flooding or breaching of water-retaining structures.
Projected impacts	• By 2050, there is an 80% chance that the average winter temperature will rise by between 0.9 °C and 2.3 °C and that the sea level will be 15 cm to 35 cm higher than in 1990. • Increase in the likelihood of flood due to an increase in sea level, as well as an increase in peak discharges from rivers in the winter (very likely). Increase in flooding in rural areas during the winter (very likely); more frequent flooding in urban areas (likely) as heavier summer storms may exceed the capacity of sewage systems designed to cope with less violent downpours. • Increase in precipitation (and decreasing contribution of snow) in winter will contribute to higher discharges in the flood basin of the Rhine and Meuse. • Increase in freshwater demand in summer due to higher temperatures and evaporation (very likely). • Greater penetration of saline water into surface water bodies (very likely). Potential salination of groundwater resources (likelihood unknown). • Decrease in levels of surface water and groundwater in the summer. • Insufficient water quality, especially due to non-point source pollution. • Longer periods of drought. • Existing unique ecosystems will be under threat. Increase in salination also constitutes a threat to existing species. However, changes also offer opportunities for new species.

Primary concerns	Water quantity	Water quality	Water supply and sanitation	Extreme weather events	Ecosystems
	✓			✓ (floods, droughts)	

Key vulnerabilities	• The Netherlands is a low-lying country situated in the Delta of the rivers Rhine, Ijssel and Meuse with around 24% of the land below sea level. Without water defences, sixty per cent of Dutch territory is vulnerable to flooding from either the sea or rivers. • The adaptive capacity of the freshwater supply is limited in the current setting. Further warming and an increasing deficit of precipitation could cause considerable problems as early as 2050.

Sources: Ministry of Housing, Spatial Planning and the Environment (2009), *Fifth Netherlands' National Communication under the UNFCCC, http://unfccc.int/national_reports/annex_i_natcom/submitted_natcom/items/4903.php* (accessed 20 June 2012); Ministry of Infrastructure and the Environment (2011), *Water Management in the Netherlands, www.rijkswaterstaat.nl/en/images/Water%20Management%20in%20 the%20Netherlands_tcm224-303503.pdf* (accessed 12 July 2012; Ministry of Transport, Public Works and Water Management; Ministry of Transport, Public Works and Water Management; Ministry of Agriculture, Nature and Food Quality; Ministry of Economic Affairs; Association of Provincial Authorities; Association of Netherlands Municipalities; Association of Water Boards(2007), *Make Room for the Climate*, Memorandum for policy discussion, *www.climateresearchnetherlands.nl/gfx_content/documents/documentation/ARK_make_ room_for_climate.pdf* (accessed 9 July 2012).

Key policy documents

Document	Reference to water?	Type	Year	Responsible institution
Delta Act[1]	Y	Legal act	2012	The Dutch Cabinet/ The Delta Programme Commissioner
National Adaptation Strategy	Y	National adaptation strategy	2007	Ministry of Transport, Public Works and Water Management
Delta Programme	Y	National adaptation action plan	2011	The Delta Programme Commissioner
Royal Netherlands Meteorological Institute (KNMI) Climate Scenarios for 2050 and 2100	Y	National impact assessment	2006	Ministry of Infrastructure and the Environment
6 Delta Area based sub-programmes	Y	Sub-national responses	Under development	The Delta Programme Commissioner

1. The Delta Act is an amendment of the Water Act.

Policy instruments

Areas \ Policy mix	Regulatory instruments	Economic instruments	Information and other instruments
Water quantity			• National Adaptation Agenda: A part of ARK (National programme for Spatial Adaptation to Climate Change), the National Adaptation Agenda sets out activities that must be undertaken to climate proof spatial planning both in theory and practice. • The Helpdesk Water: Provides an information base for people working in water policy, water management and water safety-issues. It was created through collaboration between the Dutch government, provinces, municipalities and the local water board's union, *www.helpdeskwater.nl/algemene-onderdelen/serviceblok/english*. • Living with Water: Aims to stimulate co-operation between water management and spatial planning, science and practice, economy and sociology. It is organised by the Water Knowledge Platform, which co-ordinates the supply and demand of water-related knowledge, *www.levenmetwater.nl/home* (in Dutch).
Water quality			
Water supply and sanitation			
Extreme weather events	• Room for the River: Recognising that extremely high river discharges for the Rhine tributaries will occur more frequently in the future, a package of 39 measures[1] was approved in 2007. The measures aim to ensure that the rivers can discharge the greater volumes of water forecasted without flooding. EUR 2.1 billion has set aside for the period 2008-20, *www.ruimtevoorderivier.nl/meta-navigatie/english/room-for-the-river-programme*. • Meuse Projects: Aim to improve flood protection by increasing the peak discharge level that the Meuse can handle by establishing a link between water and spatial planning. EUR 400 million has set aside for the 2008-20 period. • Climate Buffers: Serve to reduce the risk of flooding by temporary storage and thus simultaneously reducing the effects of prolonged drought. Work is underway at 35 sites, with interesting combinations of wet and robust nature, improved water security, and different functions (walking, living, water storage) per landscape type, *www.klimaatbuffers.nl/english-homepage-2*.	• Taxes paid by inhabitants and landowners of dike rings to the water boards to maintain the flood defences and regional water systems.	
Ecosystems			

1. The measures aim to meet the 2015 target for the statutory level of protection for a river discharge of 16 000 m^3/s and enhance the environmental quality of the river region. The extra room the rivers will need in the coming decades to cope with higher discharges due to the forecast climate changes will remain permanently available.

Main research programmes

- Deltares research on climate change: Deltares helps to develop new concepts for flood and drought management as well as adaptive strategies for spatial planning and the development of infrastructure and ecosystems. The scope of activities includes: Preparation of water scenarios, impact and vulnerability assessment, adaptation strategies and measures and mitigation, *www.deltares.nl/en/expertise/100795/climate-change*.
- Knowledge for Climate: Research programme for the development of knowledge and services, which makes it possible to climate proof the Netherlands. Governmental organisations (central government, provinces, municipalities and water boards) and businesses, actively participate in research programming through the input of additional resources, *http://knowledgeforclimate.climateresearchnetherlands.nl*.
- Climate Services: Founded in 2006, provides knowledge and information on past, current and future climate, including impacts on the water sector, *www.knmi.nl/research/climate_services*.
- Climate adaptation in the Dutch Delta strategic options for a climate-proof development of the Netherlands: Sets out the first steps in developing strategic adaptation options. The study analysed how The Netherlands could adapt to expected changes in climate in four areas: Flood protection; freshwater supplies; rural areas, ecosystems and biodiversity; and urban areas. It was undertaken by the Netherlands Environmental Assessment Agency (PBL) in 2011, *www.anpassung.net/SharedDocs/Downloads/DE/Climate_20Adaptation_20in_20the_20Dutch_20Delta,templateId= raw,property=publicationFile.pdf/Climate%20Adaptation%20in%20the%20Dutch%20Delta.pdf*.
- Routeplanner (2006): Commissioned by the National Programme for Spatial Adaptation to Climate Change (ARK).

Principal financing mechanisms and investment programmes

- The Flood Protection Programme: Reinforcing the primary defence structures along the coast, the rivers and the major delta waters that were shown not to meet the statutory standards. The budget of EUR 2.5 billion for the 2009-20 period will cover all measures to be taken.
- The Delta Fund: Covers the cost of measures and provisions for flood protection and freshwater supplies. Alongside construction and the improvement, the Delta Fund also provides funding for the management, maintenance, and operation of water works and related research. As of 2020, EUR 1 billion will be allocated by the Government for the implementation of the Delta Programme. The Delta Fund and the contribution from the water authorities ensure a secure source of financing for flood protection.
- Taxes paid by inhabitants and landowners to the Regional Water Authorities to maintain the flood defences and regional water systems.

Highlights and innovative initiatives

- **The Delta Programme:** Seeks to ensure that present and future generations are safe from water and will have sufficient freshwater in the centuries ahead. The programme takes an "adaptive delta management" approach, taking measures in the short term that will expand capacity to adapt to long-term changes and withstand extreme situations. *www.government.nl/issues/water-management/delta-programme.*
- **Room for the River:** Designed for the Rhine tributaries, this programme has three objectives:
  - by 2015, the branches of the Rhine will cope with a discharge capacity of 16 000 m^3/s of water per second without flooding;
  - the measures implemented to increase safety will also improve the overall environmental quality of the river region;
  - the extra room the rivers will need in the coming decades to cope with higher discharges due to the projected climate changes will remain permanently available, *www.ruimtevoorderivier.nl/meta-navigatie/english/room-for-the-river-programme.*

New Zealand

Climate change impacts on water systems

Observed changes and trends	• Average temperatures have increased by 0.9 °C between 1908 and 2006. • Decrease in the number of days of frost in most locations. • Drought over the summer and autumn of 2007 and 2008 affected many of New Zealand's regions, leading to a reduction in agricultural productivity. • Extreme rainfall events (24 hour duration) increased in the West and decreased in the North and East between 1930 and 2004.[1] • Sea level rose at an average of 1.6 millimetres per year during the 20th century, in line with global trends.
Projected impacts	• Increase in temperatures of 0.9 °C on average for the period 2030-49 and 2.1 °C for the period 2080-99, compared to the period 1980-99. The most significant increase in temperature is expected during summer and autumn and in the North Island and in the northwest of South Island. Temperature rise is projected to be the least significant during spring. • Increase in annual mean precipitation in the West (up to 5% by 2040 and 10% by 2090) and decrease in the East and North (exceeding 5% in places by 2090). Increase in annual mean rainfall is expected in Tasman, West Coast, Otago, Southland, and Chathams; decrease is expected in Northland, Auckland, Gisborne and Hawke's Bay. • Lower river flows in summer and higher flows in winter are likely. • Longer summers with higher temperatures and lower rainfall will reduce soil moisture and groundwater supplies as well as heighten water demand. • Lower river flows in summer will raise water temperatures and aggravate water quality problems. • Heavier and more frequent extreme rainfall, especially in areas where mean rainfall is projected to increase.

Primary concerns	Water quantity	Water quality	Water supply and sanitation	Extreme weather events	Ecosystems
				✓ (floods, droughts)	

Key vulnerabilities	• Agriculture and forestry are important elements of the New Zealand economy and a key export industry. Both industries are climate dependant.

1. This is most likely to be the result of natural decadal variability, although a contribution from climate change cannot be ruled out.

Sources: Ministry for the Environment (2008), *Climate Change Effects and Impact Assessment*, 2nd Edition, *www.mfe.govt.nz/publications/climate/climate-change-effect-impacts-assessments-may08/climate-change-effect-impacts-assessment-may08.pdf* (accessed 9 July 2012); New Zealand Government (2009), *New Zealand's Fifth National Communication under the UNFCCC*, *http://unfccc.int/national_reports/annex_i_natcom/submitted_natcom/items/4903.php* (accessed 22 June 2012).

Key policy documents

Document	Reference to water?	Type	Year	Responsible institution
Resource Management Act (RMA) 1991[1]	Y	Legal act	1991	Ministry of Environment, Local governments are responsible for implementation
Local Government Act 2002[2]		Legal act	2002	Local Governments
Civil Defence and Emergency Management Act 2002		Legal act	2002	Ministry of Civil Defence and Emergency Management, Local government

Key policy documents *(cont.)*

Document	Reference to water?	Type	Year	Responsible institution
Building Act 2004		Legal act	2004	Ministry of Business, Innovation and Employment (Building and Housing), Local Government

1. To provide greater legal certainty for councils about their responsibility to consider the effects of climate change, the RMA was amended in 2004 to require all persons exercising duties and functions under it to have particular regard to the effects of climate change. The New Zealand Government commenced its water reform process in mid-2009. This process is aimed at addressing the risks to the economy and the environment posed by poor management of our freshwater resource and capitalising on opportunities for economic growth that this resource provides. Reform addressing the management of freshwater takes place within broader reforms to the Resource Management Act 1991. Any proposals for legislative amendment would likely be encompassed within broader reforms to the resource management system in 2013.
2. Under the Local Government Act 2002, local governments are responsible for a range of functions that may be affected by climate change.

Policy instruments

<table>
<tr><th>Areas \ Policy mix</th><th>Regulatory instruments</th><th>Economic instruments</th><th>Information and other instruments</th></tr>
<tr><td>Water quantity</td><td>• Resource Management Act (RMA): Under the RMA, regional councils must give effect to the National Policy Statement for Freshwater Management 2011 (the "NPS"), www.mfe.govt.nz/publications/rma/nps-freshwater-management-2011.
• The NPS requires all regional councils to set objectives and limits on quality and water quantity by ensuring that their regional plans have regard to the reasonably foreseeable impacts of climate change.[1] Regional Councils are at various stages in implementing NPS requirements.
• Regional policy statements, water plans, water consents.</td><td></td><td rowspan="5">• National Policy Statement for Freshwater Management 2011 Implementation Guide (the "Guide"): A non-statutory document that provides guidance to regional councils on how to give effect to the requirement to incorporate impacts of climate change in their regional plans,[2] www.mfe.govt.nz/publications/rma/nps-freshwater-guide-2011.
• Information and guidance on climate change adaptation: The Government provides technical manuals, summary publications and guidance to inform local government, businesses and individuals about how they could be impacted by climate change and what they can do to adapt, www.climatechange.govt.nz/physical-impacts-and-adaptation/publications.html.
• Map of regional climate impacts: Provides an indication of the potential regional impacts of climate change in New Zealand. Refer to: www.mfe.govt.nz/issues/climate/about/climate-change-affect-regions.
• Sustainable Land Management and Climate Change Programme ("SLMACC"): The Ministry of Primary Industries is working with sector organisations, local government and Māori to ensure that farmers, growers and foresters have the information and tools they need to be prepared for a changing climate. A changing climate presents both challenges and opportunities for land-based sectors. The programme funds research and technology transfer, www.maf.govt.nz/environment-natural-resources/climate-change/impacts-and-adapting-to-climate-change.aspx.</td></tr>
<tr><td>Water quality</td><td>• RMA, NPS, regional policy statements, water plans, water consents.</td><td></td></tr>
<tr><td>Water supply and sanitation</td><td>• Local Government Act regulations, NPS, health standards.</td><td></td></tr>
<tr><td>Extreme weather events</td><td>• National Civil Defence Emergency Management plan, local emergency management plans, Rural Support Trusts.</td><td></td></tr>
<tr><td>Ecosystems</td><td></td><td></td></tr>
</table>

Policy instruments (cont.)

Areas \ Policy mix	Regulatory instruments	Economic instruments	Information and other instruments
			• Local government initiatives: – *www.waikatoregion.govt.nz/tr201137*; – *www.trc.govt.nz/climate-change*; – *www.boprc.govt.nz/Sustainable-Communities/Climate-Change.aspx*; – *http://ecan.govt.nz/publications/Plans/ClimateChangeReport.pdf*; – *www.wellington.govt.nz/services/environment/climate/climatechange.html*; – *www.kapiticoast.govt.nz/Your-Council/A--Z-Council-Services-and-Facilities/Stormwater*. • Rural Support Trusts: Help people and families in the wider rural community who experience an adverse event (e.g. climatic, financial or personal) to more effectively meet and overcome these challenges. Services are free and confidential, *www.rural-support.org.nz*.

1. See: Policies A1 and B1.
2. The Guide states that the starting point should be guidance documents for local government on climate change published by the Ministry for the Environment, *www.mfe.govt.nz/publications/climate*.

Main research programmes

- Guidance manuals include:
  - *Climate Change Effects and Impacts Assessment: A Guidance manual for Local Government in New Zealand*: Provides information on climate variability and observed changes in the New Zealand climate, *www.mfe.govt.nz/publications/climate/climate-change-effect-impacts-assessments-may08*.
  - *Coastal Hazards and Climate Change: A Guidance Manual for Local Government in New Zealand*, July 2008, *www.mfe.govt.nz/publications/climate/coastal-hazards-climate-change-guidance-manual*.
  - *Tools for Estimating the Effects of Climate Change on Flood Flow: A Guidance Manual for Local Government in New Zealand*, May 2010, *www.mfe.govt.nz/publications/climate/climate-change-effects-on-flood-flow/index.html*.
- The SLMACC programme funds research and technology transfer, including aspects related to water and agriculture and forestry, *www.maf.govt.nz/environment-natural-resources/climate-change/impacts-and-adapting-to-climate-change.aspx*.
- The National Institute of Water and Atmospheric Research (NIWA) undertake research into climate-related impacts on water, *www.niwa.co.nz/our-science/climate/research-projects*.

Principal financing mechanisms and investment programmes

- Government departmental funding and local government funding.
- Funding to NIWA to undertake scientific research into climate impacts.
- The Community Environment Fund, administered by the Ministry for the Environment provides funding for community initiatives that contribute to reduced greenhouse gas emissions and improved freshwater and coastal management, *www.mfe.govt.nz/withyou/funding/community-environment-fund/criteria.html*.
- The Fresh Start for Fresh Water Clean-up Fund provides funding to local councils, Iwi, and other partnership organisations, *www.mfe.govt.nz/issues/water/freshwater/fresh-start-for-fresh-water/cleanup-fund.html*.
- The Irrigation Acceleration Fund administered by the Ministry for Primary Industries supports regional-scale rural water infrastructure projects, *www.mpi.govt.nz/environment-natural-resources/funding-programmes/irrigation-acceleration-fund*.

Norway

Climate change impacts on water systems

Observed changes and trends

- Annual mean temperature for mainland Norway has increased by about 0.8 °C over the past hundred years.
- Annual precipitation has increased by slightly less than 20% since 1900. Much of the precipitation increase has occurred after 1980.
- The observed temperature increase has generally resulted in increased stream flow during winter and spring.
- There has been a trend towards earlier snowmelt, resulting earlier spring floods in recent years. Southeastern Norway has had longer periods with low stream flow in the summer. At the same time, floods from heavy rainfall have become more frequent since 1987.
- Shorter snow season in most locations during the 20th century.
- Rapid warming of permafrost in the Norwegian alpine areas. Temperature measurements performed since 1999 show the rate of warming at around 0.3 °C per decade at a depth of 25 meters.

Projected impacts

- Warming in all parts of Norway and during all seasons. Annual mean temperature is estimated to increase by 3.4 °C. Temperature increase in the northern part of Norway could be 5.4 °C towards the end of the century.
- Increase in average annual precipitation during this century by 5%, 18% and 31% by the year 2100 for low, medium and high climate projections respectively. The national average for the medium projection shows an increase in precipitation of around 20% in the autumn, winter and spring and 10% in the summer.
- Increase in annual runoff, although regional differences can be expected. In general, an increase in runoff in the autumn and winter and a reduction in runoff in most places in summer. In glacial areas, an increase in runoff is also expected in summer.
- Shorter snow season throughout the country towards the end of this century. The change will probably be greatest in lower lying areas.
- Climate change will add to the challenges that the water supply and sewerage sector are currently facing as well as pose new challenges.
- Increase in temperature and precipitation may result in more loose organic material in the water, such as pollution from agricultural activity, resulting in a change in the amount of light that penetrates lakes.
- Summer stratification period in lakes will be longer and more distinct, favouring cyanobacterial blooms.
- Flood projections are uncertain, as local variations are large. In general, floods due to rainfall can be expected to increase, whereas the probability of large snowmelt floods will be reduced. Earlier onset of spring floods, due to higher temperatures. Increase in floods in late autumn and winter. In particular, more intense local precipitation will create problems in small, steep rivers and streams and in densely populated areas.
- Reduction in stream flow and increase in soil moisture deficit due to higher temperatures and somewhat lower precipitation during the summer season, resulting in more serious summer droughts. The changes are expected to be substantial towards the end of the century, particularly in southern Norway.

Primary concerns

Water quantity	Water quality	Water supply and sanitation	Extreme weather events	Ecosystems
			✓	✓

Key vulnerabilities

- Arctic land areas have experienced more warming than any other region of the earth over the last 20 to 30 years. The climate changes seen in the Arctic have already led to major impacts on the environment and on economic activities. If the climate warming continues as projected, these impacts are likely to increase.
- Changed distribution of freshwater species, which can spread towards areas at high altitude and latitude. Lakes in these areas are particularly sensitive to variations in climate and species respond rapidly to changes in the ice regime. Some species of fish and crustaceans may be unable to migrate to alternative habitats due to isolation between freshwater systems.
- Water temperature can rise above critical levels for important fish species like salmon, trout and charr in some areas, with rivers in southern Norway and regulated rivers with low minimum water flow at highest risk.
- Increased levels of particulate organic matter will decrease light conditions in lakes and along with changes in vertical stratification in lakes will affect phytoplankton and other organisms.
- Increased flooding during autumn and winter and in small water streams can increase pollution.
- Extreme weather conditions increase risk for avalanches and landslides.
- Increased need for maintenance and improvement of water and sanitation systems. Old and under dimensioned water and drainage systems are put under pressure when precipitation increases.
- Outdoor recreation and tourism related to freshwater and fishing may need adaption.
- Due to extreme weather events, floods in smaller river streams are expected to be a more important challenge in the future.

Sources: Norwegian Ministry of the Environment (2010), *Adapting to a Changing Climate*, Official Norwegian Reports NOU 2010:10, *www.regjeringen.no/pages/36782608/PDFS/NOU201020100010000EN_PDFS.pdf* (accessed 10 July 2012); Norwegian Ministry of the Environment (2009) *Norway's Fifth National Communication under the Framework Convention on Climate Change,http://unfccc.int/national_reports/annex_i_natcom/submitted_natcom/items/4903.php* (accessed 21 June 2012).

Key policy documents

Document	Reference to water?	Type	Year	Responsible institution
National Adaptation Programme		National adaptation programme	2009	Ministry of the Environment
Official Norwegian Report (NOU) on Climate Change Adaptation[1]		National vulnerability assessment	2010	Expert Committee
White Paper on flood and landslides (Meld. St. 15, 2011-12)			2012	Ministry of Petroleum and Energy
White Paper on Climate Adaptation in Norway (Meld. St. 33, 2012-13)		National adaptation strategy[2]	2013	Ministry of the Environment

1. Included a background study by Hanssen-Bauer et al. (2009), "Climate in Norway 2100", in order to provide a joint scientific basis for assessments of vulnerability and adaptation needs.
2. Will be discussed in Parliament on 17 June 2013.

Policy instruments

Areas \ Policy mix	Regulatory instruments	Economic instruments	Information and other instruments
Water quantity	• Water Regulation: Implements the EU Water Framework Directive. • Planning and Building Act. • Nature Diversity Act. • Pollution Control Act. • Water Resources Act.		• Clearing house for climate change adaptation: Climate Adaptation Norway (Klimatilpasning Norge): Launched in 2009, this portal facilitates exchange of information and experience on climate change adaptation. Developed to meet the specific needs of regional and local spatial planners, it also serves a wider audience – researchers, businesses, administrative bodies, *www.regjeringen.no/en/dep/md/kampanjer/engelsk-forside-for-klimatilpasning.html?id=539980.* • Maps of climate in Norway in 2050 and 2100: Show future changes in temperature and precipitation in Norway in 2050 and 2100. The maps can be used as a tool in the municipalities' planning for future climate changes (e.g. risk and vulnerability analyses, spatial planning, etc.), *www.regjeringen.no/en/dep/md/kampanjer/engelsk-forside-for-klimatilpasning/temperature-and-precipitation-changes-in.html?id=609105.* • "Cities of the Future": Collaboration between the Government, Norway's 13 largest cities and three private sector organisations for business and finance to reduce GHG emission and adapt to current and future climate change. The programme runs from 2008-14,[1] *www.regjeringen.no/en/sub/framtidensbyer/cities-of-the-future.html?id=548028.* • The national mapping of flood and landslide risks takes account of expected climate change effects. The results are used in the spatial planning at district level. Climate change effects are also taken into account in the dimensioning of flood and landslide measures.
Water quality	• Water Regulation: Implements the EU Water Framework Directive. • Planning and Building Act. • Nature Diversity Act. • Pollution Control Act. • Water Resources Act.		
Water supply and sanitation			
Extreme weather events	• Planning and Building Act. • Pollution Control Act. • Water Resources Act.		
Ecosystems	• Water Regulation: Implements the EU Water Framework Directive. • Planning and Building Act. • Nature Diversity Act. • Pollution Control Act. • Water Resources Act.		

1. See additional information under "Highlights and innovative initiatives".

Main research programmes

- Programme on climate change and impacts in Norway (NORKLIMA): A 10 year research programme directed by the Norwegian Research Council. It aims to generates new knowledge about the climate system, about climate tends in the past, present and future, and about the direct and indirect impacts of climate change on the natural environment and society, to provide a basis for informing adaptive responses, *www.forskningsradet.no/prognett-norklima/Home_page/1226993599851*.
- NorClim project: Provides information on future climate development to governmental bodies, decision and policy makers, researchers, businesses, NGOs and the general public. The project involves a large number of scientists from leading research institutes, *www.norclim.no* (in Norwegian).
- Report on "Hydrological projections for floods in Norway under a future climate": Published by the Norwegian Water Resources and Energy Directorate in 2011, *www.nve.no/Global/Publikasjoner/Publikasjoner%202011/Report%202011/report5-11.pdf*.

Principal financing mechanisms and investment programmes

Highlights and innovative initiatives

- **"Cities of the Future"** is a collaboration between the Government, the business sector and the 13 largest cities in Norway .The purpose of the programme is to help and encourage the cities: To reduce their greenhouse gas emissions through urban planning and housing development, to counteract negative effects of climate change and to improve the urban environment. The program is divided in four focused areas: Land use and transport; energy in buildings; consumption and waste and adaptation to climate change. National and local government officials and politicians meet regularly to discuss how to meet challenges separately and together, share experiences and examples, develop new policies and strategies based on lessons learned and develop common knowledge and methods. The cities present common political statements to national policy processes such as the National transportation plan, the white paper of climate policies, white paper of waste management, and the white paper on climate change mitigation. The programme demonstrates that the dialog between the national and local authorities is necessary in order to get a more goal-oriented development for policy instruments in different sectors. At the same time, it is important to have a coherent city development where environmental, social and cultural initiatives work together to achieve good results.
- **"The Midgard Snake"** project of Oslo Water and Wastewater Department. Increased urban development, together with the increasing precipitation due to climate change has led to more pressure on the water mains and increased risks of flooding and water damage. "The Midgard Snake" will function as an interruptive drainage system, preventing polluted water from reaching the Oslo Fjord. This tunnel (with a capacity of 50 000 m^3) will function as both a transport route and a retention reservoir, where water can be stored if the Purifying Plant lacks capacity. The project will improve water quality in the Fjord, address climate change impacts and reduce energy consumption because the water is not being transported as far as it was previously. Estimated at NOK 1 billion, the project will be finished in 2013-14, *www.regjeringen.no/en/sub/framtidensbyer/the-participating-cities-/oslo/the-midgard-snake-in-oslo-.html?id=574174*.

Poland

Climate change impacts on water systems

Observed changes and trends	• The last two decades of the 20th century and the first decade of the 21st century were the warmest period in the entire 230 years of instrumental climate observations. • Total annual precipitation increased to 635 mm during the period 2001-05, as compared to 616 mm in the period 1971-2000. Regional differences also increased. • In recent years, abnormally warm springs (2002, 2007), autumns (2006) and winters (2000, 2006) were observed.
Projected impacts	• Significant increase in temperature by the end of the 21st century. Increase in the minimum temperature in winter, with a more pronounced increase in maximum temperature in summer. • Longer summer with frequent heat waves and droughts. • Increase of water temperature of rivers and lakes in summer months and decrease of frequency of ice formation. • Shifts in temperature and the hydrological regime may worsen water quality, especially for lowland rivers already burdened with substantial pollution.

	Water quantity	Water quality	Water supply and sanitation	Extreme weather events	Ecosystems
Primary concerns	✓	✓	✓	✓ (droughts and floods)	✓

Key vulnerabilities	• Water shortage during dry seasons and increased flood risks due to extreme weather events mainly impacting on agriculture, biodiversity, Nature2000 and urbanised areas.

Sources: Climate-Adapt, European Climate Adaptation Platform (2012), *Poland*, *http://climate-adapt.eea.europa.eu/countries/poland* (accessed 15 September 2012); Ministry of Environment (2010), *Fifth National Communication under the UNFCCC*, *http://unfccc.int/national_reports/annex_i_natcom/submitted_natcom/items/4903.php* (accessed 20 June 2012).

Key policy documents

Document	Reference to water?	Type	Year	Responsible institution
2030 National Water Management Strategy[1]	Y	National water strategy	2005	National Water Management Authority
Medium-term and Long-term Development Strategy		National development strategy	Planned for 2012	
National Strategy for Adaptation to Climate Change	Y	National adaptation strategy	Planned for end of 2013	Ministry of the Environment
		National adaptation plan	Under development	
		National risk assessment	Under development	

1. The Strategy was elaborated by the Minister of the Environment. The document was approved by the Council of Ministers on 13 September 2005.

Policy instruments

Areas \ Policy mix	Regulatory instruments	Economic instruments	Information and other instruments
Water quantity			
Water quality	• Water Law Act of 18 July 2001.[1]		
Water supply and sanitation			
Extreme weather events	• Law Act of 19 September 2011 on specific measures concerning reducing flood impacts.		
Ecosystems	• Water Law Act of 18 July 2001		

1. Dz. U. 2001.115.1229; Dz. U. 2012.145.951.

Main research programmes

• Project KLIMAT, *http://klimat.imgw.pl* (in Polish).

Principal financing mechanisms and investment programmes

Portugal

Climate change impacts on water systems

Observed changes and trends	• Overall decrease in annual runoff. • Increase in the difference between regions in terms of water availability. • Decrease in surface water quality due to temperature increase as well as non-point source pollution resulting from increased rainfall. • Decrease in groundwater quality (e.g. saline contamination). • Increase in flood risk due to increased rainfall intensity.
Projected impacts	• Increase in temperature of 3 °C to 7 °C for the summer season in continental Portugal, affecting in particular inland Northern and Central regions. Temperature increase is estimated to be more moderate in the islands, in the order of 1 °C to 2 °C in Azores and 2 °C to 3 °C in Madeira. • Reduction in annual rainfall on the continent by 20% to 40%, relative to current levels due to a shorter rainy season. • Progressive reduction in river runoff and aquifer recharge over the course of the century. By 2100, annual mean runoff shifts in various basins include: Reduction by 15 to 30% in the Vouga and Mondego Basins; shift between +5% and –10% north of River Duoro; reduction of 10 to 30% in the Tejo River Basin. • Reduction in the availability of water supplies. • Water quality degradation due to higher water temperatures and reduced river flow in the summer, particularly in the South. • Seawater intrusion into groundwater. • Increase in the magnitude and frequency of flood, especially in the North due to the concentration of precipitation in winter and the estimated general increase in the frequency of heavy precipitation events. • Increased frequency and intensity of heat waves and increased drought risk. • Displacement of wetlands.

	Water quantity	Water quality	Water supply and sanitation	Extreme weather events	Ecosystems
Primary concerns	✓			✓ (floods and droughts)	

Key vulnerabilities	• Water shortage during dry seasons in the south and increased flood risks in the North.

Sources: Portuguese Environment Agency (2010), *Fifth National Communication to the UNFCCC on Climate Change, http://unfccc.int/national_reports/annex_i_natcom/submitted_natcom/items/4903.php* (accessed 20 June 2012); Santos, F.D. et al. (2001) *Climate Change in Portugal: Scenarios, Impacts, and Adaptation Measures: Executive Summary and Conclusions*, Project SIAM, *www.siam.fc.ul.pt/SIAMExecutiveSummary.pdf* (accessed 9 July 2012).

Key policy documents

Document	Reference to water?	Type	Year	Responsible institution
National Climate Change Adaptation Strategy[1] (ENAAC)	Y	National adaptation strategy	2010	Executive Committee of the National Commission for Climate Change
National Adaptation Strategy to the Impacts of Climate Change in Water Resources (ENAAC-RH)	Y	National water sector adaptation strategy	Under development	National Water Authority (INAG)
		National adaptation plan	Under development	
National Water Plan	Y	National water sector adaptation plan	2010	INAG
Climate Change In Portugal: Scenarios, Impacts, and Adaptation Measures (SIAM) I and II Project	Y	National risk assessment	1999-2002 2002-06	Centre for Climate Change Impacts Adaptation and Modelling (CCIAM)
Regional Strategy for Climate Change[2]		Sub-national responses		

1. Resolution of the Council of Ministers No. 24/2010.
2. Government Resolution No. 123/2011, 19 October 2011.

Policy instruments

Areas \ Policy mix	Regulatory instruments	Economic instruments	Information and other instruments
Water quantity			• National Plan for Efficient Use of Water (PNUEA): Approved in 2005, promotes the efficient use of water, particularly in the urban, agricultural and industrial sectors and to help minimise the risk of water shortages and improve environmental conditions in aquatic environments. • River Basin Management Plans: The main instruments for water management, defining for each water body the quality status to be achieved in the short and medium term, as well as the programmes of measures to achieve those objectives. • Strategic Plan for Water Supply and Wastewater Treatment. • Flood Risk Management Plans.
Water quality			
Water supply and sanitation			
Extreme weather events			
Ecosystems			

Main research programmes

- Climate Change in Portugal: Scenarios, Impacts, and Adaptation Measures (SIAM): Initiated in 1999, SIAM was the most comprehensive and integrated assessment on the impacts and vulnerability associated to climate change in Portugal. The second phase of SIAM project (SIAM II) began in 2002. It focused on the Sado estuary and included the Azores and Madeira.

Principal financing mechanisms and investment programmes

- Portugal received funding through the European Economic Area (EEA)/Norwegian financial mechanisms to fund the first phase of its work on adaptation.
- Portuguese Carbon Fund (PtCF): A review of legislation for the PtCF may consider channelling some of these funds to adaptation. The PtCF is government's financial instrument to acquire Kyoto Protocol emission credits to ensure Portugal fulfils its Protocol commitments. The fund has a EUR 354 million endowment up to 2012.

Slovak Republic

Climate change impacts on water systems

Observed changes and trends	• Average annual temperature increased by 1.6 °C and annual precipitation totals decreased by 3.4% on average during the period 1881-2008. In the South, the decrease in precipitation was more than 10%; in the North and Northeast, the increase was up to 3%. • Evidence of gradual desertification, particularly in the South.
Projected impacts	• Increase in temperature of 4 °C. • Significant increase in precipitation totals by the end of the century, mainly in the form of rainfall. Increase in precipitation totals in winter (by 30% in the north and in the highlands). • Shifting snow conditions. A decrease in snow is expected at altitudes below 800 m. Above 1 200 m, temperature increases are not expected to diminish snow cover. On the contrary, peak snow cover (new snow cover and total snow cover) can be expected more often, increasing the risk of avalanches in mountainous regions. • Significant shifts (both increases and decreases) in both the temporal and territorial distribution of runoff. • Increase in runoff will also take the form of floods. Summer drought may also be interrupted by extreme floods. Winter floods will be more frequent. • Prolongation of drought periods and the increase in precipitation totals during short periods with cyclonal weather in summer. • Shifts in hydrological regimes will affect water supplies and hydroelectric power production.

	Water quantity	Water quality	Water supply and sanitation	Extreme weather events	Ecosystems
Primary concerns				✓ (floods and droughts)	
Key vulnerabilities					

Source: Ministry of Environment (2009), *Fifth National Communication of the Slovak Republic on Climate Change, http://unfccc.int/national_reports/annex_i_natcom/submitted_natcom/items/4903.php* (accessed 20 June 2012).

Key policy documents

Document	Reference to water?	Type	Year	Responsible institution
Water Act	Y	Legal act	2004	
National Adaptation Strategy		National adaptation strategy	Under development	Ministry of Environment
Water Management Policy to 2015	Y	National water strategy		
Water plan of Slovakia	Y	Water management plan		
Flood Risk Management Concept	Y	National flood risk strategy		
		National impact assessment	Under development	

Policy instruments

Areas / Policy mix	Regulatory instruments	Economic instruments	Information and other instruments
Water quantity	• The Plan of the Development of Public Water Supplies and Public Sewage Systems: Requires the re-evaluation of usable groundwater reserves with the goal of achieving good status (ecological, chemical and quantitative), taking into account climate change impacts. • The National Plan of River Basin Management: Requires taking climate change impacts into account in the re-evaluation of future water needs as well as the safety of existing water withdrawals from reservoirs for water supply and energy production and low flow augmentation.		
Water quality			
Water supply and sanitation			
Extreme weather events	• Landscape Revitalisation and Integrated River Basin Management Programme for the Slovak Republic:[1] Addresses flood prevention and protection. Local authorities have built water-holding landscapes to prevent floods and droughts in hundreds of villages and towns. • The National Plan of River Basin Management: Requires the re-evaluation of flood design requirements and a re-assessment of the safety of dams and water structures, taking into account climate change impacts.		
Ecosystems			

1. Government resolution No. 744/2010, approved on 27 October 2010.

Main research programmes

- National Climate Programme of the Slovak Republic (1994-2010): Focussed on climate trends and projects as well as climate change scenarios and impacts. It was funded by the Ministry of Environment.
- "Climate Change Impacts and Possible Adaptation Measures in Various Sectors in Slovakia": Project co-ordinated by the Slovak Hydro meteorological Institute to study projections of climate change impacts, undertake analysis of possible impacts of GDP, and propose adaptation measures.

Principal financing mechanisms and investment programmes

- Project to build a system of prevention flood-protection measures to reduce flood risk for 24 towns under the Landscape Revitalisation and Integrated River Basin Management Programme for the Slovak Republic. It was financed through grants of approximately EUR 580 000 from the Prime Minister's budgetary reserves.

Highlights and innovative initiatives

- **Landscape Revitalisation and Integrated River Basin Management Programme** for the Slovak Republic: Addresses flood prevention and protection. Local authorities have built water-holding landscapes to prevent floods and droughts in hundreds of villages and towns.

Slovenia

Climate change impacts on water systems

Observed changes and trends	• Increase in average temperature by approximately 0.4 °C per decade during the period 1971-2010 in most parts of the country. A higher increase was observed in larger towns and a smaller increase in the coastal and alpine region. • Significant inter-annual variation in precipitation. The years 1971, 1983, 2003 and 2011 were extremely dry, while 1965 was extremely wet. • Slightly significant trend of the decreasing precipitation in most of the country. On a seasonal basis, there is generally decreasing precipitation in the summer and increasing precipitation in autumn. During the winter and spring the general decreasing trend is not very significant. • Changes in river discharge regimes and river temperature regimes observed in the last decades. Observations include a significant decreasing trend in the amount of annual river discharge across the country, a general increase of extreme discharge and a seasonal change of the main discharge extreme. • Increase in the river temperature that affects the river temperature regime. • An increasing trend in number of episodes with high water level above the warning threshold has been observed. In particular, flash floods seem to have become more frequent. • Decreasing duration of lake ice on the alpine Lake Bohinj. • Longer periods of drought at the end of winter and in spring. However, summer droughts are much more problematic due to faster evaporation.[1] In past two decades there were 5 major agricultural drought events recorded (1992, 1993, 2000, 2001 and 2003) according to surface water balance. In years 2000, 2001 and 2003, drought has caused between 60% and 80% of the damage due to natural hazards. Droughts are becoming more likely to occur and the frequency of severe droughts is increasing. The last recorded severe droughts occurred in 2007 and 2012.				
Projected impacts	• Significant increase in temperature, more pronounced in the warmer half of the year. Temperatures are expected to increase the most in summer (between 3.5 °C and 8 °C), followed by winter (between 2.5 °C and 7 °C), spring (between 2.5 °C and 6 °C) and autumn (between 2.5 °C and 6 °C) by the end of the 21st century, relative to the period 1961-90. • Increase in precipitation in the winter months (up to 30%), a decrease in precipitation in the summer months (up to 20%) with no significant change in precipitation in the spring and autumn months. • Changes in precipitation will have a proportionally higher impact on river discharges. In summer, low discharges will be more pronounced.				
Primary concerns	Water quantity	Water quality	Water supply and sanitation	Extreme weather events	Ecosystems
	✓ (decrease in annual groundwater recharge rate[2] and projected decrease in the water availability)[3]	✓ (increase in pressure on water resources during drier periods may affect the water quality)	✓ (water shortages during low water season in the Southwest and the Northeast).	✓ (increased frequency of agricultural drought, especially in the Northeast; floods in the Northwest and Southwest)[4]	
Key vulnerabilities	• During summer periods, decrease in water availability is expected in the Northeast and Southwest, increasing the pressures on water resources in these regions.				

1. There are indications of an increasing trend of days with evapotranspiration greater than 5 mm (in average 25% per decade).
2. National water-balance modelling indicates slightly decrease of total annual groundwater recharge rate in last 40 years and in the future decades. Due to the high sensitivity of local shallow aquifers to changes in renewable groundwater quantity, some public water supply systems may experience temporary shortage of drinking water.
3. Due to the decrease in summer precipitation, water availability will be of great concern in the NE and SW regions of Slovenia.
4. The frequency of agricultural drought is expected to increase due to increased variability of evapotranspiration and precipitation during warm periods.

Sources: Ministry of the Environment and Spatial Planning (2010) *Slovenia's Fifth National Communication under the UNFCCC, http://unfccc.int/national_reports/annex_i_natcom/submitted_natcom/items/4903.php* (accessed 21 June 2012); Slovenian Environment Agency (2012), *National Groundwater Recharge Assessment Report 2011*, Slovenian Environment Agency; CC-WaterS Project, *Climate Change and Impact on Water Supply*, South East Europe Transnational Co-operation Programme, *http://ccwaters.eu* (accessed 14 September 2012).

Key policy documents

Document	Reference to water?	Type	Year	Responsible institution
Water Act	Y	Legal act	2012 (last amendment)	Ministry of Agriculture and the Environment
Strategy for the Transition of Slovenia to a low carbon society by 2060	Y	National climate change strategy	draft	Ministry of Agriculture and the Environment
National adaptation action plan for Slovenia	Y	National adaptation action plan	Under development, planned for 2014	Ministry of Agriculture and the Environment
Risk assessment for Slovenia	Y	National risk assessment	Under development, planned for 2013	Ministry of Agriculture and the Environment
Preliminary Flood Risk Assessment	Y	Sub-national responses	2011	Ministry of Agriculture and the Environment
The EU Danube Region Strategy (2011)	Y	Transboundary responses	2011	Ministry of Agriculture and the Environment
Sub-regional Sava river basin Management plan (expected 2013), Strategy for the Adriatic-Ionian Macro-Region (expected 2013)	Y	Transboundary responses	Under development, planned for 2013	Ministry of Agriculture and the Environment

Policy instruments

Areas / Policy mix	Regulatory instruments	Economic instruments	Information and other instruments
Water quantity		• Payments for water rights (water use allowance), water reimbursement fee (water use refund).	• The program of measures within the River Basin Management Plan (2009-2015) foresees: *i)* the preparation of a strategy and action plan to adapt water management to climate change until 2027 at the level of river basin districts; and *ii)* taking climate change into account in the development of water use. • Drought Management Centre for South-Eastern Europe:[1] Promotes the development of tools for risk assessment of drought. The primary task of the Centre is not adaptation to climate change, but developed and applied methodologies and monitoring could also serve as a good starting point for improved water management and mitigation of drought.
Water quality		• Environmental tax for environmental pollution caused by wastewater discharge.	
Water supply and sanitation		• Sustainable and rational water use is promoted via a price for "excessive" drinking water consumption. The price of "excessive" drinking water consumption is 50% higher than the price for regular consumption.	
Extreme weather events			
Ecosystems			

1. It represents the interest of 13 countries in South-Eastern Europe. Slovenia has taken over the role of host country.

Main research programmes

- The Slovenian Environment Agency provides observations and some projections for climate change impacts in Slovenia.
- Project Climate Variability in Slovenia (2009-13) produced a report of the impact of climate variability on the aquatic environment in 2010 (only in Slovene).
- Project BOBER to upgrade of the system for monitoring and analysing the water environment in Slovenia (2009-15).
- Project AlpWaterScarce to develop water management strategies to address water scarcity in the Alps (2008-11).
- Project CC-WaterS focussed on climate change impacts on water supply (2009-11).

Principal financing mechanisms and investment programmes

- Two ongoing projects addressing flood defence are financed by the EU Cohesion Fund.

Highlights and innovative initiatives

- **Centre for eco-remediation** of the Faculty of Arts of the University of Maribor: Identifies ecosystem based approaches for adaptation and proposes concrete adaptation measures.

Spain

Climate change impacts on water systems

Observed changes and trends	• General increase in temperature over the 20th century of a much greater magnitude than the global average. Temperature rise has been more accentuated in winter. • Rainfall showed downward trend over the 20th century, especially in the south and in Canary Isles. However, the high variability of rainfall precludes more precise judgment regarding trends. • Frequency of longer droughts has increased over the last decades of the 21st century.
Projected impacts[1]	• Overall increase in annual mean temperature of approximately 1.5 °C for the period 2011-40, from 2.5 °C to 2.9 °C during 2041-70 and from 3.6 °C to 4.8 °C for the period 2071-2100. Greater increases are expected during the spring and summer. • Decrease in mean annual precipitation of 5% to 6% for the period 2011-40, from 8% to 9% for the period 2041-70, and from 9% to 17% during 2071-2100. However, some projections suggest increases in rainfall in the eastern Iberian Peninsula during the period 2011-40. Greater reductions are expected in Atlantic basins and Canary Isles, the Guadalquivir River Basin and the southern Iberian Peninsula. • Decrease in water resources mainly due to the reduction in rainfall. Decreases estimated to be 8% for the period 2011-40, from 11% to 16% for the period 2041-70 and from 14% to 28% for the period 2071-2100. However, Mediterranean areas and the north-eastern Iberian Peninsula would only have minor reductions or even increase in water resources during the 2011-40 periods. • Increase in the frequency of intense and short droughts over first decades of the 21st century. Increase in the frequency of longer droughts over the last decades of the 21st century.

Primary concerns	Water quantity	Water quality	Water supply and sanitation	Extreme weather events	Ecosystems
	✓	✓	✓	✓	✓

Key vulnerabilities	• At the end of the 21st century, the most significant reductions in water resources are expected in southern and central Spain with reductions from 30% to 50% under the A2 scenario. Under the B2 scenario, the most significant reductions are expected in the southern half of the Iberian Peninsula, in the Balearic Isles, and the western Ebro River Basin, with reductions from 20% to 30%.

1. Based on two emissions scenarios (A2 and B2).

Sources: Ministry of Agriculture, Food and Environment (2010), *Assessment of the Impact of Climate Change on Natural Water Resources*, *www.magrama.gob.es/es/cambio-climatico/temas/impactos-vulnerabilidad-y-adaptacion/plan-nacional-adaptacion-cambio-climatico/rec_hidricos.aspx* (accessed 12 July 2012); Ministry of Environment and Rural and Marine Affairs (2006), *Spanish National Climate Change Adaptation Plan* (PNACC), *www.magrama.gob.es/es/cambio-climatico/temas/impactos-vulnerabilidad-y-adaptacion/pnacc_ing_tcm7-12473.pdf* (accessed 18 July 2012); Ministry of Environment and Rural and Marine Affairs (2005), *A Preliminary General Assessment of the Impacts in Spain due to the Effects of Climate Change*, Ministry of Environment and Rural and Marine Affairs.

Key policy documents

Document	Reference to water?	Type	Year	Responsible institution
National Climate Change Adaptation Plan (PNACC)	Y	National adaptation strategy	2006	Coordination Commission of Climate Change Policies (CCPCC)/Spanish Climate Change Office (OECC), Ministry of Agriculture, Food and Environment
PNACC Work Programmes (First and Second)[1]	Y	National adaptation action plan	2006/2009	OECC
A Preliminary General Assessment of the Impacts in Spain due to the Effects of Climate Change	Y	National impact assessment	2005	Ministry of Environment
Assessment of the impact of climate change on natural water resources	Y	National water sector impact assessment	2010	Ministry of Agriculture, Food and Environment

Key policy documents (*cont.*)

Document	Reference to water?	Type	Year	Responsible institution
Climate Change Adaptation Plans or Strategies of Autonomous Communities		Sub-national responses		Autonomous Communities

1. In addition, as a follow-up to the PNACC development, two monitoring reports have been published in 2008 and 2011 by the Spanish Climate Change Office (OECC), see *www.magrama.gob.es/es/cambio-climatico/temas/impactos-vulnerabilidad-y-adaptacion/plan-nacional-adaptacion-cambio-climatico/planificacion_seguimiento.aspx* (in Spanish).

Policy instruments

Areas \ Policy mix	Regulatory instruments	Economic instruments	Information and other instruments
Water quantity	• Regulation on Hydrological Planning (Royal Decree 907/2007): Requires water plans to assess climate change impacts on water resources.[1] • Instruction on Hydrological Planning (Decision ARM/2656/2008): Requires an assessment of the impact of climate change in the inventory of water resources.[2]		• Public campaign to promote saving water: A broad public awareness campaign launched by the Ministry of Environment promotes water saving and informs the public of the impacts of different domestic actions, *www.mma.es/secciones/total/index.htm.* • Adaptation guidance for local administrations: Technical guidance produced by the Spanish Network of Cities for Climate (RECC) helps local authorities to identify impacts and vulnerability to climate change and promotes adaptation.
Water quality			
Water supply and sanitation			
Extreme weather events	• Regulation on Flood Risks Management and Assessment (Royal Decree 903/2010): The preliminary assessment of flood risks will include the impacts of climate change.[3] • Review of the Basic Construction and Design Regulations and Review of Land Planning and Land Uses to take into account increased climate risk due to climate change.	• Initiatives regarding insurance for natural disasters, including floods (e.g. promotion of insurance as a prevention instrument; adaptation of the insurance market to the better address changing climatic risk).	
Ecosystems			

1. Article 11.4.
2. Article 2.4.6.
3. Article 6.

Main research programmes

- Regional climate change projections: Official climate change projections for Spain throughout the 21st century, available in the report "Generation of Regional Climate Change Scenarios for Spain", *www.aemet.es/en/serviciosclimaticos/cambio_climat* (in Spanish). The second phase production collection of projections "Scenarios-PNACC 2012".
- Impacts and vulnerability assessments: Extensive assessments have been carried out under the PNACC, including an impact assessment for water resources, *www.magrama.gob.es/es/cambio-climatico/publicaciones/publicaciones/Libros.aspx* (in Spanish).
- Agreement between the General Directorate of Water and CEDEX (Public Works Study and Experimentation Centre), by which this public institution is entrusted to carry out a research on climate change impacts on water resources and adaptation strategies. The researched was published in 2011, *www.magrama.gob.es/es/agua/temas/planificacion-hidrologica/resumenejecutivodef7_con_web_tcm7-165447.pdf* (in Spanish).

Principal financing mechanisms and investment programmes

- One of the strategic lines of the National Scientific Research, Development and Technological Innovation Plan 2008-11 is devoted to energy and climate change. The strategic line for energy and climate change had a budget of EUR 60 million in 2010 (EUR 12 million for grants and EUR 48 million for loans). In 2011, the assigned budget amounted EUR 65 million (EUR 13 million for grants and EUR 52 million for loans). Among its aims is the promotion of climate change adaptation policies, including projects related to water.[1]
- Proyecto ARCO-UPM: *Project of R&D&I: Vulnerability, Impacts and Adaptation to Climate Change: Integrated Assessment of Agriculture, Water Resources and Coasts* (ARCO), *www.upm.es/observatorio/vi/index.jsp?pageac=actividad.jsp&id_actividad=59572* (in Spanish).

1. See Programme of Work on R + D + I (2011), *www.idi.mineco.gob.es/portal/site/MICINN/menuitem.7eeac5cd345b4f34f09dfd1001432ea0/?vgnextoid=37efacc362163210VgnVCM1000001d04140aRCRD* (in Spanish).

Highlights and innovative initiatives

- The project **"Climate Change in the Spanish Coastal Areas"** (C3E, *www.c3e.ihcantabria.com*) has been carried out by the University of Cantabria. It provides a comprehensive analysis of climate change impacts in the Spanish coastal areas, as well as a set of tools to prevent and mitigate effects of climate change that might be applied in water planning in coastal areas. Water planning in coastal areas can also be useful information for river basin planning. In fact, river flows and estuary data were taken into account in the project, *www.magrama.gob.es/es/cambio-climatico/temas/impactos-vulnerabilidad-y-adaptacion/plan-nacional-adaptacion-cambio-climatico/impactos-en-la-costa-espanola-por-efecto-del-cambio-climatico/default.aspx* (in Spanish).

Sweden

Climate change impacts on water systems

Observed changes and trends	• Relatively warm climate over the past 75 years, particularly in the 1930s and since 1987. • Annual mean temperature has been around 1 °C higher and the annual precipitation around 10% higher over the past twenty years (1991-2011) compared to the previous 30-year period (1961-90). • Seasonal patterns differ somewhat from the annual mean changes. The temperature difference relative to earlier decades has been the largest in winter. The precipitation increase has been largest in summer. Generally, seasonal changes are obvious in winter, spring and summer. Autumn temperature and precipitation are the least affected. • A general tendency to less severe spring (snowmelt) flooding and more frequent riverine floods in autumn. The annual mean runoff has not changed very much. • Since 2000, many areas of the country have affected by a number of floods.
Projected impacts	• Warming of greater magnitude and at a faster rate than global mean warming. A set of mid-range unmitigated climate change scenarios suggest temperature increase of 3 °C to 6 °C toward the end of the century, depending on the season and part of the country. • Changes in precipitation range from a few % change in southern Sweden summer to an increase of 20% to 50% in winter and also for the intermediate seasons for most of the country. Changes in summer are the most uncertain. • Increased precipitation may lead to increased hydropower production and may also affect the risks for extreme floods and thus future dam safety assessments. The largest lake within EU, Lake Vänern, and its outlet, River Göta älv, constitute a complex system, where risks for flooding and landslides may increase. • Changes in river flow into Lake Mälaren, sea level rise that affects its outflow, and the rapid economic development of the shorelines have created a complex situation with conflicting interests. The city of Stockholm is located at the outlet of the lake, which provides water supply for some 2 million people in the metropolitan area. • Reduction in the quantity of precipitation that falls as snow, except in inland Norrland, where a small increase is expected. • Over time, shortening winters lead to a declining duration in the snow cover season and thinner maximum snowpack, despite increased winter precipitation. Similarly, the ice covered season (lakes, Baltic Sea) shortens by several weeks under unmitigated scenarios. • Due to decreasing snowmelt in spring, flood risks along rivers and lakes are projected to decrease in central and northern Sweden (north of Stockholm) with the exception of the far northwest which shows an increase. Flood risks generally increase in the rain-fed rivers of southern Sweden. • Decreasing availability of water resources may occur in some parts of Sweden (e.g. the south and southeast), due to small precipitation changes compounded by increased evaporation and longer growing season, which increases evapotranspiration. • Increases in water supply on average by 5% to 25% for the whole country, but decreases occur locally. Risk of water shortage in southeastern parts of the country. • Intense precipitation is projected to increase both in winter and summer, with potential consequences for urban storm drainage systems. • Sea level rise is a main concern for the coastal cities in southern Sweden and may increase risk of saltwater penetration into wells close to the coast. • Changes to the water balance (amount, timing), water temperature and related biological activity and nutrient transport may affect water quality and aquatic ecosystems.

Primary concerns	Water quantity	Water quality	Water supply and sanitation	Extreme weather events	Ecosystems
				✓ (floods)	
Key vulnerabilities					

Sources: Andréasson, J., H. Gustavsson and S. Bergström (2011), "Projekt Slussen – Förslag till ny reglering av Mälaren. SMHI Rapport nr 2011-64", Norrköping, *www.stockholm.se/Fristaende-webbplatser/Fackforvaltningssajter/Exploateringskontoret/NyaSlussen/Om-projektet/Formella-dokument/Miljodomsansokan*; Helsinki Commission (HELCOM) (2007), *Climate Change in the Baltic Sea Area*, *www.helcom.fi/stc/files/Publications/Proceedings/bsep111.pdf* (accessed 12 August 2012); Ministry of Environment (2009), *Sweden's Fifth National Communication on Climate Change*, *http://unfccc.int/national_reports/annex_i_natcom/submitted_natcom/items/4903.php* (accessed 20 June 2012); Swedish Commission on Climate and Vulnerability (2007), "Sweden Facing Climate Change – Threats and Opportunities", *Swedish Government Official Reports*, SOU 2007:60, *www.regeringen.se/sb/d/108/a/94595* (accessed 15 August 2012); Swedish Meteorological and Hydrological Institute (SMHI) (2012), www.smhi.se/kunskapsbanken/klimat/klimatindikatorer-1.7050 (climate indicators); *www.smhi.se/klimatdata/meteorologi/temperatur/1.2430* (temperature data); *www.smhi.se/klimatdata/meteorologi/nederbord/1.2887* (precipitation); *www.smhi.se/kunskapsbanken/klimat/sveriges-klimat-har-blivit-varmare-och-blotare-1.21614* (observed changes 2010-1991 vs. 1961-90); *www.smhi.se/klimatdata/klimatscenarier/klimatanalyser* (regional climate projection analyses).

Key policy documents

Document	Reference to water?	Type	Year	Responsible institution
Sweden facing climate change	Y	National impact assessment	2007	Swedish Commission on Climate and Vulnerability

Policy instruments

<table>
<tr><th>Areas \ Policy mix</th><th>Regulatory instruments</th><th>Economic instruments</th><th>Information and other instruments</th></tr>
<tr><td>Water quantity</td><td></td><td></td><td rowspan="5">• Guidance on risks and adaptation planning: Aims to provide clearer and stronger consideration given to the risk of accidents, flooding and erosion. Methods have been developed for adapting planning and construction to prevent, avoid and minimise the adverse consequences of climate change.
• Recommendations on flood risk in municipal planning. In Central Sweden, seven county administrative boards have issued joint recommendations on how to account for flood risk in municipal spatial planning.
• National portal for adaptation: www.klimatanpassning.se (in Swedish).</td></tr>
<tr><td>Water quality</td><td></td><td></td></tr>
<tr><td>Water supply and sanitation</td><td></td><td></td></tr>
<tr><td>Extreme weather events</td><td>• Amendment of the Planning and Building Act (2008): Requires that buildings may only be erected at suitable places and account has to be taken of the risk of accidents, flooding, and erosion in municipal comprehensive plans and detailed development plans.
• Climate change is considered in the new edition of the Swedish guidelines for flood design standards for dams.</td><td></td></tr>
<tr><td>Ecosystems</td><td>• Work is in progress on a new outlet structure and a new regulation scheme for Lake Mälaren, which both account for the effects of climate change.</td><td></td></tr>
</table>

Main research programmes

Principal financing mechanisms and investment programmes

- The Budget Bill in 2008 and in the 2009 Climate and Energy Bill have provided resources to intensify the work on climate adaptation. A sum of SEK 400 million has been earmarked for climate adaptation measures for 2009-11 and additional resources are proposed for 2012.

Switzerland

Climate change impacts on water systems

Observed changes and trends	• More than 50% of the volume of Swiss glaciers has been lost since 1850. During the last decades, a dramatic acceleration of the glacial melting has been observed. • Since 1965, river water temperatures increased significantly. • Two thirds of all communities have experienced flooding in the last 30 years.
Projected impacts	• Overall, ample precipitation, the balancing effect of the snow cover and melting water from glaciers are expected to ensure sufficient quantity of water resources. • Yet, with the possible decrease in average summer rainfall and the number of rainy days in combination with rising summer temperatures, extremely dry periods might last longer and occur more frequently. • Higher river water temperatures imply that habitats for cold water fish will shrink and warm water fish will have a larger natural environment. • A rising snow line brings about the potential for more frequent floods, landslides, etc., during winter. • The intensity and frequency of extreme events and natural hazards, including heavy precipitation, floods, slope instabilities, landslides, rock falls, heat waves and droughts are expected to increase, but in most cases the signals of the models are weak.

Primary concerns	Water quantity	Water quality	Water supply and sanitation	Extreme weather events	Ecosystems
		✓ (during low flow periods)		✓	✓ (conflict management between conservation and utilisation for low flow periods)

Key vulnerabilities	• Swiss Alpine regions are particularly vulnerable.

Sources: Advisory Body on Climate Change (2007), *Climate Change and Switzerland 2050: Expected Impacts on Environment, Society and Economy*, ProClim – Forum for Climate and Global Change, Forum of the Swiss Academy of Sciences, *http://proclimweb.scnat.ch/portal/ressources/794.pdf* (accessed 11 August 2012); CCHydro (2012), *Effects of Climate Change on Water Resources and Watercourses, www.bafu.admin.ch/publikationen/publikation/01670/index.html?lang=en* (accessed 15 August 2012); Swiss Confederation (2009), *Switzerland's Fifth National Communication under the UNFCCC, http://unfccc.int/national_reports/annex_i_natcom/submitted_natcom/items/4903.php* (accessed 20 June 2012).

Key policy documents

Document	Reference to water?	Type	Year	Responsible institution
CO2 Act[1]		Legal act	Revised in 2011	Federal Council
CO2 Ordinance		Legal act	Under development	Federal Council
Water Act (L'eaux)[2]	Y	Legal act	Revised in 2011	Federal Council
Water Ordinance	Y	Legal act	Revised in 2011	Federal Council
Adaptation to climate change in Switzerland, objectives, challenges and areas of action: The first part of the strategy of the Federal Council	Y	National adaptation strategy	2012	Federal Council
	Y	National adaptation action plan	Planned for 2013	Federal Council
Swiss Strategy on Natural Hazards	Y	Natural hazards strategy		PLANAT

Key policy documents (*cont.*)

Document	Reference to water?	Type	Year	Responsible institution
Updated assessment of climate change-related risks and opportunities	Y	National risk assessment	Under development	
Climate Change and Switzerland 2050	Y	National impact and vulnerability assessment	2007	Advisory Body on Climate Change (OcCC)
Regional level adaptation documents in several cantons		Sub-national responses	2011-ongoing	Cantons
Adaptation to climate change in Swiss cities		Sub-national responses	2011-12	FOEN

1. The new CO_2 Act, approved by the Swiss parliament on 23 December 2011 sets out the legal framework for Switzerland's climate policy from 2013 to 2020. While the previous version of the CO_2 Act (entered in force in May 2000) focussed exclusively on mitigation, the revised law also provides the legal basis for adaptation.
2. The revised law aims to restore the natural functions of water bodies with an obligation for cantons to stipulate the space provided for surface water and the obligation to plan and implement restoration measures. Water Act: *www.admin.ch/ch/e/rs/c814_20.html*; Ordinance: *www.admin.ch/ch/e/rs/c814_201.html*.

Policy instruments

Areas \ Policy mix	Regulatory instruments	Economic instruments	Information and other instruments
Water quantity	• Review of the existing legislation with regard to minimum low flows, lake regulation and management, and multi-purpose use of alpine reservoirs).		• Information portal on climate change adaptation, *www.bafu.admin.ch/klimaanpassung/index.html?lang=fr* (in French). • Guidelines and recommendations to identify risk areas and to implement a water resource management/river basin management. • Guidelines and promotion for: Strategic planning, interconnection of water supplies, and regionalisation of waste water treatment plants.
Water quality	• Review of the existing legislation with regard to sewage disposal and discharge of cooling water.		
Water supply and sanitation			
Extreme weather events		• Mandatory insurance schemes for natural disasters for all real estate owners.	
Ecosystems	• Modification of Water Law (2011): Promotes restoration of rivers and wetlands to allow rivers to recover their structure and services and allow a better adaptation to climate change.		

Main research programmes

- The research programme CCHydro Effects of climate change on water resources and watercourses is examining the impacts of climate change on the hydrological cycle, *www.bafu.admin.ch/publikationen/publikation/01670/index.html?lang=en* (in English).
- Research project assessing the impact of climate change on hydropower production in the Swiss Alps, *www.hydrologie.unibe.ch/projekte/Synthesebericht.pdf* (in German), *www.hydrologie.unibe.ch/projekte/Rapport%20de%20synthese.pdf* (in French).
- Impacts 2014: Assessment of climate change impacts in Switzerland (to be published in 2014).
- National Research Program 61 (NRP61) on "Sustainable Water Management": Aims to determine the capacity of natural systems to absorb the effects of changes in environmental conditions (climatic conditions) in order to develop intelligent and forward-looking strategies for sustainable and integral water resources management, *www.snf.ch/E/targetedresearch/researchprogrammes/currentNRP/Pages/_xc_nfp61.aspx* (in English).

Principal financing mechanisms and investment programmes

- Yet to be addressed, but expected to be covered in Part 2 of the Adaptation Action Plan (the second part of the strategy of the Federal Council, in preparation).

Turkey

Climate change impacts on water systems

Observed changes and trends	• Prominent increase in summer temperature across the country, particularly in the west and southwest. Winter temperatures have shown a general decreasing trend in the last five decades. • Significant decrease in winter precipitation in the western provinces over the last fifty years. An increase in fall precipitation in northern parts of the Central Anatolia. • Sea level rise of around 12 cm over the last century for the Mediterranean and Black Sea regions.
Projected impacts	• Increase in the annual mean temperature for Turkey is around 2 °C to 3 °C for the period 2071-2100 relative to the period 1961-90. In winter, the projected temperature increase is higher in the eastern half of the country in winter. In summer, this pattern is reserved and the western half of the country, especially the Aegean region, will experience temperature increases up to 6 °C. • Increasing temperatures would lead to increased summer temperatures, reduced winter precipitation (especially in the western provinces), loss of surface waters, more frequent arid seasons, degradation of soil, erosion in coastal regions and floods all of which are direct threats to water resources. • Precipitation decreases along the Aegean and Mediterranean coasts; the most severe reductions will be on the southwestern coast. Increases in precipitation along the Black Sea coast; the Caucasian coastal region is expected to receive substantially more precipitation. Central Anatolia shows little or no change in precipitation. • Reduction of up to 200 mm of snow water equivalent over the high plains of eastern Anatolia and the eastern part of the Black Sea. This may result in major changes in the stream flow of Turkey's river basins. • Shift from snowfall to rainfall during winter by the year 2100, due to increasing temperatures. • Continued sea level rise, with possibility of saltwater intrusion. • Eutrophication and salination of shallow lakes and wetland constitute an ecological problem, but may also become an economic burden, since it may not be possible to use water with toxic algae explosions or saline water for drinking or irrigation purposes. • Increase in the frequency, intensity and duration of extreme weather events such as drought in the South, Southeast and West and flood especially in the Western Black Sea region. • Increase in water temperature will affect ecological processes, geographic distribution of aquatic species resulting in extinction of species and loss of biodiversity.

Primary concerns	Water quantity	Water quality	Water supply and sanitation	Extreme weather events	Ecosystems
	✓	✓	✓	✓ (flood and drought)	✓

Key vulnerabilities	• Possible major changes in the quantity and quality of water and stream flow in Turkey's river basins. Rivers are the main sources of water for Turkey, not only for safe drinking water, domestic and industrial use, but also for irrigation and power generation. • Water resources needed for food production and rural development are threatened by climate change impacts such as an increase in summer temperatures, a decrease in winter precipitation (in western provinces in particular), a loss of surface waters, an increase in the frequency of droughts, land degradation, coastal erosion and floods. • Key economic sectors, enterprises and populations will be negatively affected by droughts and floods.

Sources: Ministry of Environment and Forestry (2007), *First National Communication on Climate Change*, General Directorate of Environmental Management, Ministry of Environment and Forestry, Ankara; Ministry of Environment and Urbanization (2011), *Turkey's National Climate Change Adaptation Strategy and Action Plan* (Draft), Ministry of Environment and Urbanization, General Directorate of Environmental Management, Department of Climate Change, Ankara.

Key policy documents

Document	Reference to water?	Type	Year	Responsible institution
National Climate Change Adaptation Strategy and Action Plan	Y	National adaptation strategy and plan	2012	Ministry of Environment and Urbanisation

Policy instruments

Areas / Policy mix	Regulatory instruments	Economic instruments	Information and other instruments
Water quantity	• Prevention of illicit use of groundwater resources (2012 onwards).	• Identification of economic instruments to ensure effective and efficient use of water (2011-15). • Incentives to promote private investments in irrigation (construction and operation of facilities) (2012-14).	• Revision of institutional and sectoral strategy plans (industry, agriculture, energy, tourism, urban, drinking water) of organisations involved in water management to include the scope of combating climate change (2011-13). • Reduce water losses in the agriculture sector (2011-15). • Improvement of local capacity of water user organisations and irrigation businesses to take into account the impacts of climate change as part of the Agricultural Drought Combat Strategy and Action Plan (2011-14). • Promotion of manufacture and use of household and industrial equipment with low water consumption. Support of projects aimed to increase recycling of process and cooling waters in priority sectors with high water consumption. Increasing pilot implementations (2014-20).[1] • Preparation of guidance for water efficiency in industry and promoting pilot practices (2012-14). • Innovative solution models to increase adaptation capacity, innovative solutions will be formulated, developed and disseminated (2011-15). • Raising public awareness of illicit use of groundwater resources (2012 onwards). • Ensuring integrated water management and planning in settlements (2011-23).[2] • Risk maps for floods, landslides, etc., to inform risk management processes and land use planning (2011-15). • Establishment and dissemination of monitoring, forecast and early warning systems for natural disasters caused by climate change (2011-13). • Raising public awareness (2011-20) and developing the capacities of relevant agencies and organisations (covering all administrative levels at the local level) with regard to risk mitigation, emergency response and post-disaster short and long term recovery approaches and practices (2011-15). • Conducting R&D studies to identify and monitor the effects of climate change on inland water ecosystems (2012-15).
Water quality	• Review of the Law No. 5686 on Geothermal Resources and Natural Mineral Waters in view of climate change impacts to prevent the loss of surface water quality (2011-15).	• Economic instruments to promote the treatment of waste water for its use in agriculture and industry (2011-20).	
Water supply and sanitation		• Development of a pricing policy to increase efficient water use in cities (2011-20).	
Extreme weather events	• Development and enforcement of legislation on the structural effects of natural disasters caused by climate change (2011-15).	• Promotion of the use of private and public insurance mechanisms to address risks from natural disasters. Studies will be undertaken (2013-15).	
Ecosystems	• Develop legislation to protect ecosystems and identify the natural structures that reduce the impacts of natural disasters occurring due to climate change (2013-15).		

1. Research and Development support provided by the Ministry of Science, Industry and Technology.
2. Wastewater Treatment Action Plan (2008-12), Potable and Industrial Water Supply Action Plan for 81 Provinces (2008-12).

Main research programmes

- Dokuz Eylul University Water Resources Management Research and Application Centre have undertaken modelling studies to investigate the likely consequences of possible global climate change on watershed scale. Pilot case studies have focussed on two major river basins located in western Anatolia, along the Aegean coast – the Gediz River and the Buyuk Menderes Basins. The Gediz is characterised by water scarcity, mainly due to competition for water among various users (mainly irrigation and rapidly growing industrial demand). It is already fully allocated, based on current demand, with no reserves for further allocation. The Buyuk Menderes is one of the longest rivers in the Aegean region and forms the Buyuk Menderes Delta, recognised as a RAMSAR site. The basin is also the main producer of cotton in Turkey and home to 2.5 million people in three major cities. The studies of climate change impacts on the basins indicated that nearly 20% of surface water will be reduced by the year 2030. By 2050 and 2100, the reduction is expected to rise to nearly 35% and greater than 50%, respectively. The decreasing surface water potential of the basins will cause serious water stress problems among users.
- Istanbul Technical University – "Impact of Climate Change on the Euphrates River Flows" (supported by TUBITAK) underway since 2010.
- Istanbul Water and Sewerage Administration (ISKI) – "Future of Climate Change Impacts on Water Resources in Istanbul and Turkey Project".
- The General Directorate of State Hydraulic Works (DSI) 6th Regional Directorate – "Identification of Surface Water Resources Potential and Flood Risks within the Perspective of Developing Water Resources Management Policies in Seyhan Basin within the Framework of Adaptation to Climate Change Project".
- Several studies to ensure adaptation to the impacts of climate change in water resources management are set out in the Action Plan:
  - Studies on hydrological drought (2014-20).
  - Research and evaluations to integrate the impacts of climate change into water resources planning activities (2011-15).
  - Projections of sectoral water demand in basins, taking into account climate scenarios (2011-20).
  - Assessment of the vulnerability of river basins and sub-basins and groundwater resources to climate change along with the development and implementation of adaptation actions (2011-20).

Principal financing mechanisms and investment programmes

- Plan to create innovative and sustainable additional financing resources to support the efforts for adapting to climate change.
- Turkey is the first country to benefit from the Climate Investment Funds of the World Bank.

United Kingdom

Climate change impacts on water systems

Observed changes and trends	• Temperature has risen in Central England by about 1 °C since the 1970s, with 2006 the warmest year on record. • Annual mean precipitation over England and Wales has not changed significantly since records began in 1766. Scotland is on average 20% wetter than it was in 1961. • Seasonal rainfall is highly variable but appears to have decreased in summer and increased in winter. • All regions of the UK have experienced more winter rainfall from heavy precipitation events. In summer, all regions except Northeast England and Northern Scotland show decrease in rainfall.				
Projected impacts	• Average annual temperatures may rise by 2 °C to 3.5 °C by the 2080s. • Annual average precipitation across the country might decrease slightly, by 0% to 15% by the 2080s. However, the seasonal distribution of precipitation will change significantly. In winter, increases in precipitation projected in the range of 10% to 30 % over the majority of the country, while a reduction in precipitation is expected in summer. • Summers will continue to become hotter and drier. By 2040, average summer temperature for the UK is expected to rise by 0.5 °C to 2.0 °C, depending on the region. The largest reductions in summer precipitation (of 40 %) are expected in the far South of England. • Milder and wetter winters. By 2030, increases in precipitation of up to 30% in the winter months. • Lower water levels in rivers and lakes would reduce their capacity to dilute pollutants and worsen water quality. • Significant decrease in the amount of snowfall throughout the UK with projected decreases in winter mean snowfall of 65% to 80% over mountain areas and 80% to 95% elsewhere. • Large reductions in summer flows coupled with increasing demand could have significant consequences for water supply. • Increase in the prevalence of extreme weather events. More frequent periods of heavy rainfall leading to increased flooding, especially in winter. • Shifts in water availability and water temperature may have implications for habitats and species and therefore ecosystem services. An increase in certain invasive non-native species may affect water quality and water industry assets.				
Primary concerns	Water quantity	Water quality	Water supply and sanitation	Extreme weather events	Ecosystems
	✓		✓ (particularly in the South and East)	✓ (floods)	
Key vulnerabilities	• Maintaining water supplies during summer droughts particularly in Southeast England. • Flood risk is projected to increase significantly across the UK.[1]				

1. Annual damage to UK properties due to flooding from rivers and the sea currently totals around GBP 1.3 billion. For England and Wales alone, the figure is projected to rise to between GBP 2.1 billion and GBP 12 billion by the 2080s, based on future population growth and if no adaptive action is taken.

Sources: Department of Energy and Climate Change (2009), *The UK's Fifth National Communication under UNFCCC*, *http://unfccc.int/national_reports/annex_i_natcom/submitted_natcom/items/4903.php* (accessed 20 June 2012); Department for Environment Food and Rural Affairs (DEFRA) (2012), *Summary of the Key Findings from the UK Climate Change Risk Assessment 2012*, DEFRA, London, UK, *www.defra.gov.uk/sac/files/SAC1215-CCRA-Paper-Annex-1-Key-Findings.pdf* (accessed 10 November 2012); DEFRA (2011), *Water for Life: Market Reform Proposals*, DEFRA, London, UK, *www.gov.uk/government/publications/water-for-life-market-reform-proposals* (accessed 20 August 2012); Jenkins, G.J. et al. (2009), *UK Climate Projections: Briefing Report*, Met Office Hadley Centre, Exeter, UK; Jenkins, G.J., M.C. Perry and M.J. Prior (2008), *The Climate of the United Kingdom and Recent Trends*, Met Office Hadley Centre, Exeter, UK; Rance, J. et al. (2012), *Climate Change Risk Assessment for Water Sector*, DEFRA, London, UK.

Key policy documents

Document	Reference to water?	Type	Year	Responsible institution
Climate Change Act[1]		Legal act	2008	Department for Environment Food and Rural Affairs (DEFRA)
Adapting to Climate Change: A Framework for Action	Y	National adaptation strategy	2008	DEFRA
		National adaptation plan	Under development	
Departmental Adaptation Plans		Adaptation plans	2010	Government Departments
England (and UK reserved matters) National Adaptation Programme	Y	National adaptation programme	Planned for 2013	DEFRA
Future Water	Y	National water strategy	2008	DEFRA
Water for Life	Y	Water white paper	2011	DEFRA
UK Climate Change Risk Assessment (UKCCRA)	Y	National risk assessment	2012	DEFRA
Regional Climate Change Partnership (Climate UK)	Y	Sub-national responses		Regional agencies, UK Climate Impacts Programme (UKCIP), and DEFRA

1. The Climate Change Act 2008 gained Royal Assent on 26 November 2008. It made the UK the first country in the world to have a statutory framework for adapting to climate change.

Policy instruments

Areas \ Policy mix	Regulatory instruments	Economic instruments	Information and other instruments
Water quantity	• Statutory requirement adopted in 2007 obliging water companies to prepare and maintain water resources management plans that look ahead 25 years. The requirement aims to secure a long-term sustainable supply and demand balance for the supply of water.	• Promote greater trading of abstraction licences and bulk supplies of water to make supply system more flexible (planned).	• The Adapting to Climate Change (ACC) Programme: Ensures that policies on both adaptation and mitigation are joined up and complementary,[1] *www.ukcip.org.uk/government/central-government/acc.* • UK Climate Impacts Programme (UKCIP): At the boundary of science, policy and practice, UKCIP co-ordinates and influences research into adaptation to climate change, and encourages organisation to use tools and information to help them consider their climate risks and how to adapt, *www.ukcip.org.uk.* • Green Book Supplementary Guidance on "Accounting for the Effects of Climate Change" produced for the Treasury's Green Book, which sets out the economic guidance used by government to assess spending, investment and policy decisions, *http://archive.defra.gov.uk/environment/climate/documents/adaptation-guidance.pdf.*
Water quality			
Water supply and sanitation	• Reporting Power: The Climate Change Act 2008 introduced a new power for the Secretary of State to direct "reporting authorities" (companies with functions of a public nature, such as water utilities) to prepare reports on how they are adaptation to climate change. The Government has developed statutory guidance setting out the process that organisations need to go through to assess risks from climate change and draw up adaptation plans, *http://archive.defra.gov.uk/environment/climate/documents/interim2/report-guidance.pdf.*		

Policy instruments *(cont.)*

Areas \ Policy mix	Regulatory instruments	Economic instruments	Information and other instruments
Extreme weather events	• Flood and Water Management Act (2010): Implements Sir Michael Pitt's recommendations requiring urgent legislation, following his review of the 2007 floods. It provides for better, more comprehensive management of flood risk for people, homes, and businesses, helps safeguard community groups from unaffordable rises in surface water drainage charges and protects water supplies to the consumer, *www.defra.gov.uk/environment/flooding/legislation.* • National Flood and Coastal Erosion Risk Management Strategy (2011):[2] Sets out a statutory framework to help communities, the public sector, and other organisations to work together to manage flood and coastal erosion risk. It supports local decision making and ensures that risks are managed in a co-ordinated way across catchments, *www.environment-agency.gov.uk/research/policy/130073.aspx.* • Sustainable Drainage Systems (SUDS): A range of measures designed to mimic as closely as possible natural drainage and its advantages in providing habitat, filtering pollutants, recharging groundwater. SUDS are particularly important in water stressed areas, and in coping with heavy downpours, thus reducing flood risk.	• Statement of Principles: A voluntary agreement between the Government and the Association of British Insurers (revised in 2008) to ensure that flood risk is managed effectively and that competitively priced flood insurance remains widely available for households and small businesses. It will not be renewed after its expiry in June 2013. Therefore, effort is underway by the Government to reach an agreement with insurers whereby insurance bills remain affordable without placing unsustainable costs on wider policyholders and the taxpayer.	• Local Authority Adaptation Indicator (NI188): The UK Government introduced in 2008 an indicator on climate change adaptation in the Local Government Performance Framework. • Flood Warning Direct: Flood warning service to provide advance notice for floods. In some parts of England, floods from groundwater are also covered, *www.environment-agency.gov.uk/homeandleisure/floods/38289.aspx.*
Ecosystems			

1. The ACC Programme in England is in two phases. Phase 1 (2008-11) will lay the groundwork necessary to implement Phase 2 – a statutory National Adaptation Programme (NAP), as required by the Climate Change Act. Defra is responsible for developing the NAP, which will be published in 2013, to address the risks set out in the first UK Climate Change Risk Assessment.
2. The Flood and Water Management Act 2010 requires the Environment Agency to "develop, maintain, apply and monitor a strategy for flood and coastal erosion risk management in England".

Main research programmes

- UK Climate Projections (UKCP09): Provide climate information designed to inform planning for climate change adaptation. UKCP09 is the fifth generation of climate change information for the UK and presents probabilities of different future climates. These probabilities are created by weighting future climate projections on how well they represent the past climate, so they can be seen as the relative degree to which each climate outcome is supported by the evidence available, *http://ukclimateprojections.defra.gov.uk.*
- UK Climate Impacts Programme (UKCIP): See further details below, *www.ukcip.org.uk.*

Principal financing mechanisms and investment programmes

- Investments for flooding and coastal erosion: Defra expects to spend at least GBP 2.17 billion on flooding and coastal erosion over the next four years 2012-15, an average of GBP 544 million per year. This sum consists of around GBP 1.04 billion in capital investments and around GBP 1.13 billion on other programme and administrative costs, such as maintenance, flood forecasting, and incident response. Defra remains committed to fully funding local authority new burdens under the Flood and Water Management Act. Up to GBP 36 million a year will be provided directly to lead local flood authorities. Defra expected that as a result of this investment, deliver better flood protection will be delivered to 145 000 households by March 2015.
- Real Options Analysis for the Thames Estuary 2100 Project: The Environment Agency's Thames Estuary 2100 (TE2100) project is developing a strategy for tidal flood risk management to the year 2100. The Thames estuary floodplain contains 1.25 million people (one sixth of London's population), about GBP 200 billion of property, and key transport and infrastructure assets, including the London Underground, 16 hospitals and eight power stations. Given the value of assets at risk, the long lead times involved in developing solutions and the uncertainty of future climate effects and the potential for learning, a flexible, adaptive approach to incorporating climate change has been taken. The project identified options to cope with different levels of sea level rise, and the thresholds at which they will be required. The options were designed to implement the small incremental changes common to all options first, leaving major irreversible decisions as far as possible into the future to make best use of the information available. The strategy can be reappraised in light of the new information and options can be brought forward (or put back). See the TE2100 case study in the Supplementary Green Book Guidance (2009), "Accounting for the Effects of Climate Change", *http://archive.defra.gov.uk/environment/climate/documents/adaptation-guidance.pdf.*

Highlights and innovative initiatives

- **UK Climate Impacts Programme (UKCIP):** At the boundary of science, policy and practice, UKCIP co-ordinates and influences research into adaptation to climate change, and encourages organisation to use tools and information to help them consider their climate risks and how to adapt. The range of tools includes, for example, the "Adaptation Wizard" to assess vulnerability, identify key climate risks and development and adaptation strategy. A methodology for costing the impacts of climate change and comparing these to costs of adaptation measure is also available, *www.ukcip.org.uk.*
- **Green Book Supplementary Guidance:** Green Book Supplementary Guidance on "Accounting for the Effects of Climate Change". New supplementary guidance on adaptation has been produced for the Treasury's Green Book, which sets out the economic guidance used by government to assess spending, investment and policy decisions. The guidance sets out the criteria that determine when it is particularly important to consider the risks and effects of climate change if a programme, policy or project. It provides tools for climate change risk assessment and offers real options analysis as an options appraisal framework, able to incorporate the uncertainty of climate change and the value of flexibility into decision making, *http://archive.defra.gov.uk/environment/climate/documents/adaptation-guidance.pdf.*

United States

Climate change impacts on water systems[1]

Observed changes and trends	Water is an issue in every region. Examples of regional changes already observed include: • Over the past several decades, extended dry periods have become more frequent in parts of the United States, especially in the southwest and the eastern US. Important agricultural areas in the central US are currently experiencing severe to exceptional drought conditions. • Reduction of snowpack/snow-water equivalent in the West. Over the last 50 years, there have been widespread temperature-related reductions in snowpack in the West, with the largest reductions occurring in lower elevation mountains in the Northwest, and in California, where snowfall occurs at temperatures close to freezing point. The Northeast has also experienced snow pack reductions over a similar period. • Increase in the incidence of heavy precipitation events across most of the US. In the past century, averaged over the US, total precipitation has increased by about 7%, while the heaviest 1% of rain events increased by nearly 20%. This has been particularly noteworthy in the Northeast. • Increase in stream flow in the eastern US. During the last century, consistent increases in precipitation have been found in the Midwest and the Northeast, along with the increased runoff. • Reduced ice cover on the Great Lakes.
Projected impacts	• All of the impacts describes above are projected to continue, along with additional emerging disruptions to the current state of the water cycle. • Temperature increases are expected to change the mix of precipitation toward more rain and less snow. Such precipitation shifts affect the origin and timing of runoff, leading to less runoff from spring snowmelt and more runoff from winter rainfall, particularly in high-latitude or mountainous areas. • Northern areas will receive more precipitation, and southern areas, particularly in the West, will become drier. • Precipitation changes have led to concerns that both droughts and floods will occur more frequently and be more severe. Increased frequency and intensity of the heaviest downpours. • In the Western US, less total annual rainfall, less snowpack in the mountains, and earlier snowmelt mean that less water will likely be available during the summer months when demand is highest. • Warmer air and water temperatures along with changes in precipitation patterns exacerbate water pollution problems, which could lead to an increase in the number of water bodies categorised as "impaired". • Increase in the frequency, severity and duration of drought, changing patterns of precipitation and snowmelt, increased evaporation, and aquifer saltwater intrusion impact on the availability of drinking water supplies. • Heavy downpours increase the amount of runoff into rivers and lakes, washing sediment, nutrients, pollutants, trash, animal waste, and other materials into water supplies, making them unusable, unsafe, or in need of water treatment. • Warmer air and water temperature and changing flows will result in deterioration of aquatic ecosystem health in some areas. • A warmer climate in the Southwest is predicted to result in widespread tree mortality that likely will cause substantial changes in forest and species distributions. These changes will alter watershed hydrology in differing ways.

Primary concerns	Water quantity	Water quality	Water supply and sanitation	Extreme weather events	Ecosystems
	✓	✓	✓	✓ (drought, especially in the West)	

Key vulnerabilities	• Many areas, particularly in the Southwest, are likely to suffer from conflict over water resources by 2025, even in the absence of climate change owing to a combination of factors including population trends and potential endangered species' need for water. • Impacts on coastal areas resulting from a combination of sea level rise, increased damage from floods and storms, coastal erosion, salt water intrusion to drinking water supplies, as well as increasing temperature and acidification of the oceans.

1. The next National Climate Assessment (NCA), conducted under the auspices of the Global Change Research Act of 1990, is scheduled to be completed in 2013. The NCAs act as status reports about climate change science and impacts. Relevant chapters for freshwater systems include: Chapter 3 – Water Resources; Chapter 10 – Water, Energy, and Land Use; Chapter 11 – Urban Systems, Infrastructure, and Vulnerability; and Chapter 25 – Coastal Zone Development and Ecosystems.

Sources: Brekke, L.D. et al. (2009), *Climate Change and Water Resources Management*, Federal Perspective, Circular 1331, US Department of the Interior and US Geological Survey, Reston, Virginia, US; Interagency Climate Change Adaptation Task Force (2011), *National Action Plan: Priorities for Managing Freshwater Resources in a Changing Climate*, *www.whitehouse.gov/sites/default/files/microsites/ceq/2011_national_action_plan.pdf* (accessed 10 August 2012); US Department of State (2010), *Fifth National Communication of the United States of America under the UNFCCC*, *http://unfccc.int/national_reports/annex_i_natcom/submitted_natcom/items/4903.php* (accessed 20 June 2012); US Drought Monitor (2012), *www.droughtmonitor.unl.edu* (accessed 14 November 2012); US Environmental Protection Agency (2012), *National Water Program 2012 Strategy: Response to Climate Change*, Environmental Protection Agency, Washington, DC; US Global Change Research Programme (2009), *Global Climate Change Impacts in the United States*, Cambridge University Press, New York, *http://oceanservice.noaa.gov/education/pd/climate/teachingclimate/climate_impacts_report.pdf* (accessed 15 July 2012).

Key policy documents

Document	Reference to water	Type	Year	Responsible institution
Executive Order 13514	Y	Executive order	2009	US Federal Agencies
SECURE Water Act[1]	Y	Legal act	2009	US Department of the Interior, Bureau of Reclamation
Biggert-Waters Flood Insurance Reform Act	Y	Legal act	2012	Federal Emergency Management Agency
Interagency Climate Change Adaptation Task Force Progress Reports	Y	Progress reports on climate change adaptation	2010, 2011	Interagency Climate Change Adaptation Task Force
Draft National Ocean Policy Implementation Plan	Y	Implementation plan (draft)	2012	National Ocean Council
Climate Change and Water Resources Management	Y	Interagency report	2009	Department of Interior
National Water Programme 2012 Strategy: Response to Climate Change		National water strategy	2012 (public comment draft)	US Environmental Protection Agency
National Action Plan: Priorities for Managing Freshwater Resources in a Changing Climate	Y	National adaptation plan for freshwater	2011	Interagency Climate Change Adaptation Task Force
Global Climate Change Impacts in the United States	Y	National risk assessment	2009	The US Global Change Research Programme (USGCRP)

1. The Act authorised the Bureau of Reclamation to continually evaluate and report on the risks and impacts from a changing climate and to identify appropriate adaptation and mitigation strategies utilising the best available science in conjunction with stakeholders, *www.usbr.gov/climate/SECURE*.

Policy instruments

Areas \ Policy mix	Regulatory instruments	Economic instruments	Information and other instruments
Water quantity			• Climate Ready Water Utilities Initiative: developed by the US Environmental Protection Agency (EPA) to assist water and wastewater utilities in becoming "climate ready". It supports the implementation of plans and adaptation strategies at water and wastewater utilities that account for potential climate change impacts and build water sector resilience, *http://water.epa.gov/infrastructure/watersecurity/climate*. • Flood: Risk Mapping, Assessment, and Planning (MAP), *www.fema.gov/rm-main#1*. • WaterSMART (Sustain and Manage America's Resources for Tomorrow) Programme: The SECURE Water Act authorises federal water and science agencies to work together with state and local water managers to plan for climate change and the other threats to water supplies and to take action to secure water resources for the communities, economies and the ecosystem they support. To meet these challenges, Water SMART was established in 2010, *www.usbr.gov/WaterSMART/index.cfm*.
Water quality	• Water quality standards for surface waters, *http://water.epa.gov/scitech/swguidance/standards/index.cfm*.		

Policy instruments (*cont.*)

Areas \ Policy mix	Regulatory instruments	Economic instruments	Information and other instruments
Water supply and sanitation	• EO 13514 requires federal agencies to reduce potable water consumption intensity 26% by FY 2020 (compared to a FY 2007 baseline). This extends the water consumption intensity reduction requirement of EO 13423 by five years.		• Healthy Watershed Initiative: The EPA, in partnership with others, launched the Healthy Watersheds Initiative to protect and maintain healthy watersheds with natural, intact aquatic ecosystems, prevent them from becoming impaired, and accelerate restoration. Initiative includes both assessment and management approaches that encourage states, local governments, watershed organisations, and others to take a strategic, systems approach, *http://water.epa.gov/polwaste/nps/watershed/index.cfm*. • WaterSense: A partnership program of the EPA that seeks to protect the future of the nation's water supply by offering people a simple way to use less water with water-efficient products, new homes, and services, *www.epa.gov/watersense*.
Extreme weather events	• Stormwater management: under the new Section 438 of the Energy Independence and Security Act (EISA) of 2007, federal agencies have new requirements to reduce stormwater runoff from federal development and redevelopment projects in order to protect water resources. The EPA issued technical guidance on the implementation of stormwater management for federal projects, *www.epa.gov/owow/NPS/lid/section438*.	• Biggert-Waters Flood Insurance Reform Act of 2012: The recent reform of the National Flood Insurance Programme (NFIP) introduced a number of measures synergistic with adapting the programme to climate change, *www.fema.gov/national-flood-insurance-program*.	
Ecosystems			

Main research programmes

- US Global Change Research Programme: Co-ordinates and integrates federal research on changes in the global environment and their implications for society, *www.globalchange.gov*.
- EPA's Water Resource Adaptation Program (WRAP): Scientists and engineers investigate the potential effects of climate change on the nation's watersheds and water infrastructures. Based on the results of these investigations, practical and effective adaptation solutions are being developed. The program's research is also a part of the *Drinking Water Research Program* and the *Global Change Research Program* in the Office of Research and Development, *www.epa.gov/nrmrl/wswrd/wq/wrap/index.html*.
- Federal Climate Change and Water Working Group (CAWWG): Formed in 2008, the group is pursuing many collaborative efforts including working with the federal and non-federal water management community to identify the most critical gaps to forecast and adapt to climate change, conduct collaborative research and technology development, and engage in a dialog, in which decision-making informs climate science research priorities, *www.esrl.noaa.gov/psd/ccawwg*.
- WaterSMART Basin Study Program: Assesses climate change risks for water and environmental resources in major river basins, including the Colorado, Columbia, Klamath, Missouri, Rio Grande, Sacramento, San Joaquin, and Truckee River basins. The program is undertaken by the US Department of the Interior's Bureau of Reclamation to accomplish its authorities under the SECURE Water Act, *www.usbr.gov/WaterSMART/docs/west-wide-climate-risk-assessments.pdf*. In the context of the Basin Study Program, the Colorado River Basin Water Supply and Demand Study released in December 2012, *www.usbr.gov/newsroom/newsrelease/detail.cfm?RecordID=41645*.

Principal financing mechanisms and investment programmes

- The Water Infrastructure Resiliency and Sustainability Act of 2011: Submitted to the US Congress in October 2011. This bill would authorise the Administrator of the EPA to establish a program of awarding grants to owners or operators of water systems to increase the resiliency or adaptability of the systems to any ongoing or forecasted changes to the hydrological conditions of a region of the United States.

Highlights and innovative initiatives

- **Reform of the National Flood Insurance Programme (NFIP):** Recent reforms to the NFIP (Biggert-Waters Flood Insurance Reform Act of 2012) introduced a number of measures synergistic with adapting the programme to climate change, *www.fema.gov/national-flood-insurance-program*.
- **Agency Adaptation Plan reporting for 2012:** Required Federal agencies to take into account national cross-cutting strategies (among other things), such as the National Action Plan for Managing Freshwater Resources in a Changing Climate. Public release of the plans is forthcoming (pending final review and certification by the Office of Management and Budget).
- **Interagency Climate Change Adaptation Task Force:** Convened in 2009, the Task Force is co-chaired by the Council on Environmental Quality (CEQ), the Office of Science and Technology Policy (OSTP), and the National Oceanic and Atmospheric Administration (NOAA). It includes representatives from more than 20 Federal agencies. On October 2009, President Obama signed an Executive Order directing the Task Force to develop a report with recommendations for how the Federal Government can strengthen policies and programs to better prepare the Nation to adapt to the impacts of climate change, *www.whitehouse.gov/administration/eop/ceq/initiatives/adaptation*.

European Commission

Climate change impacts on water systems

Observed changes and trends	• Global mean temperature has increased by 0.8 °C compared with pre-industrial times. Europe has experienced greater warming than the global average. • Precipitation changes show spatially variable trends across Europe. Increase in annual precipitation in northern Europe by 10% to 40% and decrease up to 20% in some parts of southern Europe during the 20th century. • Increasing trend in annual river flows showed in northern parts of Europe over the 20th century, with increases mainly in winter, and a slightly decreasing trend in southern parts of Europe. • Significant acceleration of the melting of European glaciers since 1980. • More flooding and heavy rain events have occurred in recent years. • Several major droughts have occurred in recent decades, such as the catastrophic drought in the summer of 2003 in central parts of the continent and the 2005 drought in the Iberian Peninsula.
Projected impacts	• Increase in temperature of 1 °C to 5.5 °C by the end of the century, higher than the projected global warming of 1.8 °C to 4 °C. • Increase in mean annual precipitation in the North and decrease in the South. • Decrease in annual river flow in southern and south-eastern Europe and increase in northern Europe. However, precise changes remain uncertain. • Significant changes in the seasonality of river flows across Europe. Decrease in summer flows in most of Europe, including in regions where annual flows will increase. • Increase the frequency and severity of droughts due to river flow in southern and south-eastern Europe, the United Kingdom, France, Benelux, and western parts of Germany over the coming decades. In snow-dominated regions, where droughts typically occur in winter, river flow droughts are projected to become less severe because a lower fraction of precipitation will fall as snow in warmer winters. • Substantial decrease in glacier coverage across Europe's mountains over the coming decades. Alpine glaciers could all but disappear this century. If glaciers continue to retreat at current or even faster rates, many areas will be put at much greater risk of floods, water shortages and sea level rises. • Decreasing quantity of fresh groundwater resources, especially in coastal areas and in southern Europe, while brackish and salt groundwater bodies will expand. In addition, fresh groundwater bodies will become more vulnerable to pollution due to reduced turnover times and accelerated groundwater flow. • Higher water temperatures and extreme weather events such as flooding and droughts will also impact on water quality and exacerbate existing pollution problems. • Increase in the frequency and intensity of floods in large parts of Europe. In particular, flash and urban floods, triggered by local intense precipitation events are likely to be more frequent throughout Europe. Flood hazard will also probably increase during wetter and warmer winters, with more frequent rain and less frequent snow. Even in regions where mean river flows will drop significantly, as in the Iberian Peninsula, the projected increase in precipitation intensity and variability may cause more floods. • Increase in the frequency and intensity of droughts in many regions of Europe as a result of higher temperatures, decreased summer precipitation, and more frequent and longer dry spells. • The regions most prone to an increase in drought hazard are southern and south-eastern Europe, but minimum river flows will also decrease significantly in many other parts of the continent, especially in summer.

Primary concerns	Water quantity	Water quality	Water supply and sanitation	Extreme weather events	Ecosystems
	✓			✓	

Key vulnerabilities	• Western Europe is vulnerable to water scarcity, droughts, and floods. The energy sector is the most vulnerable. A shift in cooling systems, together with reduced thermal electricity production, can help to overcome water shortages and protect ecosystems from thermal pollution. Some minor irrigated agricultural areas may not suffer from water shortages due to increasing efficiencies. However, due to temperature increases, maize yields are expected to decline, meaning that either cropping calendars or cropping patterns need to be adapted. Due to climate change impacts and increasing future water abstractions, minimum water requirements for ecosystem maintenance and the hydropower sector are at risk. Navigation will suffer from climate change during either drought periods or flooding. The largest unknown is with regard to future water quality, which is expected to decrease resulting from diffuse source loadings released with floods and heavy rainfall or reduced dilution capacity of the rivers. Transboundary river basins are of particular interest, as they have to deal with many kinds of vulnerabilities.

Climate change impacts on water systems *(cont.)*

- In Eastern Europe, water scarcity can be reduced due to integrated water management. There is no major water user that is particularly threatened. The region is also vulnerable to floods, with the highest costs related to damages in percent of GDP. Similar to Western Europe, transboundary rivers have to deal with high risk of flooding upstream, whereas downstream vulnerability is related to water shortages and droughts. Navigation and ecosystems are threatened by climate change impacts, which will be exacerbated by increasing abstractions.
- In Southern Europe, freshwater resources will suffer in the future from climate change impacts as well as socio-economic drivers. The region is highly vulnerable to water scarcity and drought and to flash floods. The agricultural sector is the most vulnerable. Reducing water abstractions can help to address the imbalance between water supply and demand. Technological changes and raising awareness will not be sufficient to reduce water stress and reduction in irrigated areas and changes in cropping calendars or cropping patterns should be taken into consideration. Of specific interest are transboundary river basins shared between Spain and Portugal. High water abstractions upstream not only cause water shortages downstream but could also lead to deterioration of groundwater aquifers due to saltwater intrusion and reduced river discharges.

Sources: European Commission (2009), *Fifth National Communication from the European Community under the UNFCCC, http://unfccc.int/national_reports/annex_i_natcom/submitted_natcom/items/4903.php* (accessed 22 June 2012); European Commission (2012), *Adaptation to Climate Change, http://ec.europa.eu/clima/sites/change* (accessed 20 April 2013); European Commission (2009), "Climate Change and Water, Coasts and Marine Issues", *Commission staff working document*, COM(2009)147 final, European Commission, Brussels; European Environment Agency (2007), "Climate Change and Water Adaptation Issues", *Technical Report*, No. 2/2007, European Environment Agency, Office for Official Publications of the European Communities, Luxembourg; European Environment Agency, Joint Research Centre, and World Health Organization (2008), "Impacts of Europe's Changing Climate – 2008 Indicator-Based Assessment", *EEA Report*, No. 4/2008, and *JRC Reference Report*, No. JRC47756, Office for Official Publications of the European Communities, Luxembourg; Flörke et al. (2011), *ClimWatAdapt Report*, Study for the European Commission, DG Environment, Brussels.

Key policy documents[1]

Document	Reference to Water?	Type of Instrument	Year	Responsible Institution
Water Framework Directive	Y	Legal Act	2000	DG Environment
Floods Directive	Y	Legal Act	2007	DG Environment
EU Adaptation White Paper	Y	Communication	2009	DG Environment
Guidance document on adaptation to climate change in water management	Y	Guidance document	2009	DG Environment
EU Adaptation Strategy		Adaptation Strategy	2013	DG Climate Action

1. For more information see *http://ec.europa.eu/environment/water/adaptation/index_en.htm*.

Policy instruments[1]

Areas \ Policy mix	Regulatory instruments	Economic instruments	Information and other instruments
Water quantity	• Water Framework Directive (WFD): Goundwater quantitative status and surface environmental flows needed to ensure good ecological status.	• Water pricing: The WFD requires that the price charged to water users adequately covers costs, including environment and resource costs, hence taking into account vulnerability to water scarcity and droughts.	• Public participation procedures in the preparation of the River Basin Management Plans required by the WFD and Flood Risk Management Plans required under the Floods Directive. All assessments, maps and plans prepared shall be made available to the public. • Droughts Management Plans. • Guidance on natural water retention measures to increase drought resilience and reduce flood risks to be prepared by 2013 as follow-up of the 2012 Blueprint to safeguard Europe's waters, to support the forthcoming River Basin Management Plans, and Flood Risk management Plans to be presented by Member States in 2015.
Water quality	• Water Framework Directive. • Urban Waste Water Directive. • Nitrates Directive.	• Water pricing under the WFD.	
Water supply and sanitation	• Urban Waste Water Directive. • Drinking water Directive.	• Water pricing under the WFD.	
Extreme weather events	• The Floods Directive requires that flood risk maps are drawn up by 013 for river basins and associated coastal areas at risk of flooding. By 015, flood risk management plans are to be established, *http://ec.europa.eu/environment/water/flood_risk/index.htm*.	• Potential implementation of payments for ecosystem services (PES) linked to natural water retention measures aiming at floods and droughts prevention.	

Policy instruments[1] *(cont.)*

Areas \ Policy mix	Regulatory instruments	Economic instruments	Information and other instruments
Ecosystems	• Framework Directive. • Integrated Coastal Zone Management (ICZM). • Natura 2000.	• Potential implementation of PES.	• European Climate Adaptation Platform (CLIMATE-ADAPT): On-line information platform to support Europe in adapting to climate change. An initiative of the European Commission, it helps users to access and share data and information on adaptation, *http://climate-adapt.eea.europa.eu/web/guest.* • The Water Information System for Europe (WISE): An on-line platform on water-related information for government, water professionals and scientists, as well as the general public, *http://water.europa.eu.* • WFD circa – Information Exchange Platform: An on-line platform established by the Commission. Circa ("Communication Information Resource Center Administrator") promotes information exchange about the implementation of the WFD between countries, European institutions, various stakeholders, and the interested public, *http://ec.europa.eu/environment/water/water-framework/iep/index_en.htm.*

1. The instruments indicated in this table provide the framework within which the policies of the EU member states are developed.

Main research programmes

- Joint Research Centre (JRC):[1] Research in support of EC climate change policy focuses on five areas: mitigation; adaptation; scenario modeling; monitoring and verification; and civil society perspectives. In particular, JRC climate change research aims to determine costs and benefits (both in monetary and non monetary terms) of mitigation and adaptation policies.
  - European Drought Observatory (EDO): A tool for assessing, monitoring and forecasting droughts on a continental level in Europe, *http://edo.jrc.ec.europa.eu/edov2/php/index.php?id=1000.*
  - Impact assessment for the Blueprint: Developed a baseline scenario bringing together climate, land-use and socio-economic scenarios and looking at the implication for water resources availability and use under different policy scenarios. A multi-criteria optimisation of scenarios for the protection of water resources in Europe is being developed to support the impact assessment of the Blueprint (forthcoming).
- The 7th Framework Programme (FP7) for Research and Technological Development: Climate change is included under the theme "environment" (one of ten themes), which has been granted EUR 1.89 billion for the period 2007-13. Research focuses on implementation of mitigation and adaptation options, in particular technological developments. Since 2003, expenditures on climate research in the EC Framework Programme are estimated at nearly EUR 570 million. Some of the most relevant projects include:
  - REFRESH: Adaptive strategies to mitigate the impacts of climate change on European Freshwater Ecosystems, *www.refresh.ucl.ac.uk.* Development of system enabling water managers to design cost-effective restoration programmes for freshwater ecosystems at local and catchment scales, accounting for future impacts of climate and land-use changes.
  - ACQWA: Assessing Climate Change Impacts on the Quantity and Quality of Water, *www.acqwa.ch.* Assessing vulnerability of water resources in mountain regions over the next 50 years, identifying possible conflicts among economic actors (users of water resources) and assessing governance/adaptation options.
  - CLIMATEWATER (*www.climatewater.org*): Study of European and international adaptation measures and strategies and how these are taken into account in water policies in order to formulate a coherent framework.
  - CLIMATECOSTS, CLIMSAVE, MEDIATION.
- ClimWatAdapt: "Climate Adaptation – Modelling Water Scenarios and Sectoral Impacts" sheds light on both vulnerability and adaptive capacity in different sectors and across Europe's river basins. It is using an integrated assessment framework consisting of scenarios on climate change and socio-economic developments, vulnerability indicators, an inventory of measures, instruments and methods to assess the performance of adaptation options, and decision support. It will cover the EU27 Member States. (*www.climwatadapt.eu*).

1. JRC is the research arm of the European Commission. It provides scientific and technical support to the development and implementation of EC policies, and it serves the interests of the Member States as a reference centre for science and technology issues.

Principal financing mechanisms and investment programmes

- EU Structural and Cohesion Funds: Provide some means to co-finance capital-intensive investment in water infrastructure and help EU Member States comply with water legislation. For management of water resources, EUR 8 billion in total funding is provided for reducing leakage rates, connecting to water supply, generating additional supply and improving infrastructure. For disaster prevention, EUR 7 billion is available. The Solidarity Fund (EUSF) provides funds for disaster relief in member states. Around EUR 1 billion are allocated each year, *http://ec.europa.eu/regional_policy/funds/cf/index_en.htm.*
- Life + Funds: Support environmental and nature conservation projects throughout the EU. For the period 2007-13, EUR 1.7 billion is available, *http://ec.europa.eu/environment/life/funding/lifeplus.htm.*
- Revenue from auctioning allowances under the Community greenhouse gas emission allowance trading system (EU ETS). The EU White Paper *Adapting to Climate Change: Towards a European Framework for Action* (2009) supports the possibility of using such revenue for adaptation purposes. The revised Directive governing the EU ETS provides that at least 50% of the revenue generated from auctioning allowances should be used, *inter alia*, for adaptation in Member States and developing countries.

Highlights and innovative initiatives

- **European Innovation Partnership for Water:** To support and facilitate the development of innovative solutions to deal with water-related challenges, as well as to support economic growth by bringing such solutions to the market. At the same time, innovations are considered to be an important tool to develop adequate and state of the art European water policy, *http://ec.europa.eu/environment/water/innovationpartnership/index_en.htm.*

ORGANISATION FOR ECONOMIC CO-OPERATION AND DEVELOPMENT

The OECD is a unique forum where governments work together to address the economic, social and environmental challenges of globalisation. The OECD is also at the forefront of efforts to understand and to help governments respond to new developments and concerns, such as corporate governance, the information economy and the challenges of an ageing population. The Organisation provides a setting where governments can compare policy experiences, seek answers to common problems, identify good practice and work to co-ordinate domestic and international policies.

The OECD member countries are: Australia, Austria, Belgium, Canada, Chile, the Czech Republic, Denmark, Estonia, Finland, France, Germany, Greece, Hungary, Iceland, Ireland, Israel, Italy, Japan, Korea, Luxembourg, Mexico, the Netherlands, New Zealand, Norway, Poland, Portugal, the Slovak Republic, Slovenia, Spain, Sweden, Switzerland, Turkey, the United Kingdom and the United States. The European Union takes part in the work of the OECD.

OECD Publishing disseminates widely the results of the Organisation's statistics gathering and research on economic, social and environmental issues, as well as the conventions, guidelines and standards agreed by its members.

OECD PUBLISHING, 2, rue André-Pascal, 75775 PARIS CEDEX 16
(97 2013 09 1 P) ISBN 978-92-64-20043-2 – No. 60661 2013

IWA Publishing's authorised EU representative for General Product Safety Regulations is Diane D'Arras, 15 rue Duret, 75116 Paris, France, e-mail: safety@iwap.co.uk.

Printed and bound by CPI Group (UK) Ltd, Croydon, CR0 4YY
12/05/2026
02108886-0008